IDENTITY IN SPIRITUALITY

THE BOOK ABOUT SELF

DR. VAL KARANXHA

First published 2021
by Helios Publishing
Connecticut, United States.
Helios Publishing is an independent publishing house.
The right of Val Karanxha to be recognized as the author has been asserted in accordance with the Copyright Act.
Copyright © 2021
Val Karanxha
Helios Publishing
Identifiers:
ISBN 978-1-7371175-0-6 paperback

All rights reserved. This book or any portion thereof may not be reproduced or used in any manner whatsoever without the publisher's express written permission except for the use of brief quotations in a book review.

Sections: Spirituality-Philosophy-Psychology-Religion-Quantum Physics

www.heliospublishing.us

The book is printed in United States of America

About the author

Val Karanxha is an Adjunct Professor at Western Connecticut State University, where she teaches theories and principles of Bilingual Education and Second Language Acquisition. She is a five-time fellowship recipient at Yale University - Teacher Institute. Val Karanxha is a published author as well as a contributing author with Routledge Publishing. Throughout the years, she has written various articles published by Yale New Haven Teacher Institute and international journals of pedagogy. Val Karanxha is fluent in five languages. She has a Master of Arts degree in Romance Languages and a Doctorate in Leadership and Policy Studies.

Contents

INTRODUCTION 5

A SOCIAL VIEW OF SELF 24

A THEORETICAL VIEW OF THE INNER-SELF 35

THE SIPIRTUAL SELF 53

THOUGHT AND MIND 80

THE UNIVERSAL CONSCIOUSNESS 106

THE PHILOSOPHY BEHIND SPIRITUAL PRACTICE 118

CONCLUSION 145

BIBLIOGRAPHY 149

INTRODUCTION

I wrote the introduction in the "we" form because when you bought or downloaded the book, you decided to be part of this beautiful journey. I welcome you on this spiritual path, and I hope that you enjoy the book.

I live not too far from the beach and in the summertime, I go for a walk early to catch the sunrise. After I walk for approximately two miles, I usually find a quiet place where I can face the ocean and just close my eyes. One day, I was sitting at the beach, gazing at the sun rays reflecting on the water. Endless shiny droplets changed the color of the water from blue to gold. I contemplated every droplet. The feeling of eternal existence within me invaded my thoughts and my mind, and moments like this make me wonder about the universe. I sat quietly, and once I closed my eyes, I felt pressure in between the eyebrows. For those who meditate, they can relate to this experience. It is the third eye-opening, the intuition, the insight. It is the moment that your mind becomes a liaison between cosmos and our physical existence on earth. People are inclined to think individually. In our human nature - we strive to seek happiness in a smaller, individualistic scale, right here, in our own little world. Yet we tend to think about the "bigger picture" when we close our eyes and meditate. It is then, when we search for answers to find our place in the universe and make a connection with it.

The path to spirituality starts with self-identity, knowing yourself, who you are, and asking questions about your existence. It takes a simple question such as: Who am I? Why am I here? That helps us understand our very own purpose in life. We often ask ourselves the question: Are we who we say we are, or who others want us to be? It all starts with the perception of self-identity, how we perceive ourselves and how others perceive us. In a perfect world, I am my own person, in harmony with myself and the universe, without constraints and ties to bind or imprison me. It would be a perfect definition of freedom. In a given world, we perceive ourselves as perfect creations. There is good inside every one of us, no matter how the others see us.

My interest in spirituality became forthcoming when I started grad school, and I began simultaneously reading books and practicing mindfulness. Meditation until grad school was only a word that did not relate to anything or any practice in my culture. I remember years ago, when I was in Athens, Greece and I saw a young girl sitting in a yoga position on the ancient walls of the Acropolis, gazing at the sun away on the horizon. That image stuck with me for so many years. She was meditating. It wasn't until I was "enlightened" and awakened by the feeling that there is a world that we cannot perceive with our senses, but if we use the power of the mind, we will be able to experience it. How we explore and understand our self is very individual. People are constantly in a dialectic relationship with their inner selves in search for peace and happiness. The answer to this conflicting nature of self is not in the alter-ego and finding the best version of oneself, or your heroic self, but finding your true-self.

Little does it matter where the study of the universe started. In ancient scriptures of Vedas, the universe is perfect, omnipotent. Later, Greek sages and philosophers left their mark with their endless calculations to figure out the natural state of things and attribute it a philosophical meaning. The study of the "world" indeed started in ancient Greece, a place where religion, philosophy and science were in harmony. The word "phisis" in ancient Greek means the essence of things and the Greek mathematicians, physicists and philosophers believed that cosmos is an organism supported by pneuma – or the cosmic breath. [1]

The duality between rationalism and mysticism is the key to bridge modern and traditional conceptualization of our reality. [2] *Yet the common path that brings these two concepts together is crucial in schematizing them cohesively in our understanding. The parallelism between these two philosophies is scaled in the same way, yet they differ in interpretations. Self as a concept, in its totality, energy and matter, physical bodies, physics of our universe and beyond, are widely discussed and touched upon subjects and have intrigued many philosophers and different schools of thought.*

Spirituality has been practiced and studied for centuries. Religion as an institution became prevalent in the medieval era, and dominated for many reasons and purposes, through coercion and control. The growing communities built their cult institutions. People flocked, churches and temples in search to purify their soul and become closer to the Creator. But did they? For centuries churches collected wealth and power. The institutionalization of the cult is the

---

[1] Karpouzos, Alexis. *Universal Consciousness: The Bridges between Science and Spirituality*. Athens, Greece: Think.Lab, 2020.

[2] Karpouzos, Alexis. *Universal Consciousness: The Bridges between Science and Spirituality*. Athens, Greece: Think.Lab, 2020.

fundamental difference between religion and spirituality. You can never institutionalize spirituality. In other words, spirituality is a philosophy more so than religion.

In the schema held by religion for centuries, the approach was a top down mind frame where self was quite out of the equation, or the person was always a sinner. While clergy preached the gospel, "self" was thrown out of balance because people would have questioned the entire schema how clergy functioned. Systematically, people were indoctrinated with many dogmas - but the common assumption is that people are programmed not to think about "self" or put "self"[3] first. To men, the only way this exhibition of egocentric behavior or selfishness would be destroyed is if we deemed "self" as an egoistic concept and attacked it to eventually save our souls.[4] In other words, to stop being selfish. This was a blow to the concept of self, the purity or our unique souls and has been promulgated by clergy for centuries. Self was punished and never glorified. Instead prioritizing self was identified and selfishness.

Even in today's education systems, the study of self as a concept is missing in many curriculums except

---

[3] Romans 2:8 ESV - - Bible Gateway. (n.d.). Retrieved April 15, 2021, from https://www.biblegateway.com/passage

[4] James 3:16 ESV - - Bible Gateway. (n.d.). Retrieved April 15, 2021, from https://www.biblegateway.com/passage

for the study of the human body and health form a medical perspective to prevent the use of drugs, and promiscuity. Perhaps, if we try to understand self-first, we would be able to emphasize and teach the importance of our existence, and people would start to love themselves more. Hence, self is a wide and complex concept and not as easy as it appears. When dealing with self, so far, we have seen different approaches in various disciplines.

The study of self has long intrigued philosophers – starting with Socrates, Plato, Descartes, Kant and many others have been fascinated by the idea that we have a spiritual self that is immortal. Sociology and social sciences see self as a part of group identity and as such we, must always belong to community of people whether ethnic, or class stratified. The physical self – needless to say - is the very core of today's medicine. We study the human anatomy from the genes to the respiratory or circulatory system to better understand it and find the right cure for diseases. Thus, the physical self, is widely studied and researched. Self and inner self along with the cognitive processes studied by psychology are as well widely explored, as they should be. Since psychology is a science that studies behavior, different scientists have different perspectives on self that are at times conflicting – especially when it comes to cognitive processes. There is a gap, a discrepancy in religious studies and spirituality has always been concerned with meditation practices without first offering an approach about the inner self, or what self is in spirituality.

When I started to explore spirituality, I did not think of self at all. Many of the books I'd read were written by practitioners in the field. These authors had spent their lives in India's Ashrams, frequented Mount Shasta, and had been in countless retreats. I envied these people, yet the books they wrote were heavy in Sanskrit terms, and it did not resonate with me. They were not written for people who wanted to practice meditation, and I was looking for more than knowledge. I was looking for the practical side of it. Writing about the practice of Hinduism without explaining the very core of spirituality was incomplete knowledge. Sometimes, the expertise in these books lacked a context, and very often I read books with a context, yet no conceptual framework or substance in spirituality.

Later on, I expanded my reading and showed an interest in books by light workers, those who worked with angels, ascension, guided meditations, law of attraction, yoga, chakras and so forth. Therefore, I formed a solid background knowledge in the field of metaphysics and spirituality, along with the practical side of it. The more I practiced, the more I came to realize that something was missing. Practicing without a Guru, or a guidance makes things a bit challenging because you always wonder of you are doing the right thing. What I was missing was the trust in my senses, and the trust is my inner self. My very understanding of spirituality was questioned by me, and my conscience. I was always doubtful if the meditation was guided by the divine, connected with the Universal Consciousness, or if it was just a product of my subconscious mind.

Analyzing self is not as easy as it might sound: People are well aware of many "selves." We have a physical body, a subtle body we call a soul, an inner self, an emotional self, and our mind. All these bodies unite cohesively and become our identity, which makes us unique. We have a social identity used to classify and stratify us into various groups and social strata to which we belong and are aware. We also have a spiritual identity, called subtle body, inner self, or soul. Hence so far, we have two bodies and two identities: a physical body with social identity and a subtle body with a spiritual essence.

The first identity is invented for social purposes. When we come into this world, we are given a name, develop a sense of belonging to our family, society, a group of people, and realize that just by being members of that particular community. We are social beings, and as such, we must function further to advance the need for procreation, survival and wellbeing. This invented identity does not die after death. The memory lives in the family or the community for decades and even centuries. People now are starting to find their roots and where they came from. The sense of belonging is what gives life its meaning.

The second identity, as I mentioned, is subtle. That identity is purposeful to us only if we are alive and dependent on the physical body. That means that we must experience the subtle body through our mindfulness – thus, to experience the subtle identity, we need the mind as an instrument to make that connection. Not all people experience the subtle body, although everyone is capable of it. Once we become aware of our subtle existence – using precisely our mind – we can see through our subtle identity, realize who we truly are, or see our true self.

How do we understand the subtle body?

We are "children of the light," and we project light and energy – and I will explain how that can be achieved. The physical self and the subtle self are two distinct identities, and throughout this book, we will learn

13

how these identities make-up for our complete image of self.

In my previous work, I have written books in political science and published many articles with topics varying from social sciences to literature and creative writing. I thought about how to present this book to the readers and wondered who my audience will be. I have read and studied books on identity, yet identity as a theory could not explain my perspective. The social science theories have a very constructivist approach to problems. Therefore, they do not deal with spirituality to define identity, meaning they deal with group behavior, but not with their metaphysical and spiritual beliefs. Social sciences describe characteristics based on patterns; their relationship does not cover people as physical or spiritual subjects. The identity that social science explains is merely a group identity. Thus, the topic I chose to explore is more profound than identity itself. Parts of this book will cover the identity issue, yet it will go much more in-depth. When talking about self, and for that matter, inner self, the topic becomes more complex. Very few have approached spiritual identity with its merits to give people the mere sense of spiritual fulfillment.

"A clear relationship among self, mind, and body in humans is still not explored in depth in philosophy and sciences regardless their interests. Modern

philosophers tend to live in an observable reality and analyze things from their perspective. The human behavior and human mind are yet to be explored and the lack of data does not give it the opportunity to better explain or explain it objectively (Gusnard, 2009; Vacariu, 2011; Dress-Langley & Durup, 2012)[5]

Nowadays the market is inundated with hundreds and thousands of books that deal with spirituality, guided meditation, laws of universe, self-realization matter, and so forth, but few of these books take a direct approach of how the spirit impact us, our inner-self and how can we maintain a spiritual awareness while we practice daily, or whenever we have time. Hence, from what I wrote before and my field of study, this book offers some key explanation of our existence. In fact, I decided to write this book years ago when I started to study spirituality. I always thought I was not prepared enough to organize my thoughts and put them into a book and perhaps, I was right. I have wondered about various topics related to spirituality and since it is such a wide topic, I decided to cover what's important to me, and what I think will benefit the reader. What is the meaning of the inner self? Fair enough, people spend time, energy, and money taking care of their inner self through therapy,

---

[5] Chung, Sung Jang. "The Science of Self, Mind and Body." *Open Journal of Philosophy* 02, no. 03 (2012): 171-78. doi:10.4236/ojpp.2012.23026.

meditation, health coaching, and improvement of all sorts. It is paramount to understand the inner-self and without an understanding of it, there can never be inner peace, says a Buddhist concept, widely embraced by the western teaching of spirituality.

Simply said, I wanted to share my experiences with the readers, practice spirituality, help them understand what spirituality is, and how we appreciate the practical side of it. To begin, it is important to know that eighteen percent of the American people identify as spiritual and not religious. Spiritual people are not bound to any religion, yet they are very godly and practice devoutness on their own terms. Most of spiritual people practice meditation, yoga, healing and mindfulness. Being a spiritual person myself, I tried to understand the world in my own terms. But little did I know until I discovered the interpretations forwarded by quantum physics about the relationship between me, the world, and the Universal Consciousness.

My experiences have varied from daily meditation, to spinning with earth's electromagnetic field. Sounds unbelievable. But it's true, and it happened to me two or three times, twice in a home setting and once meditating outside. Approximately ten years ago, when I started to study spiritually, I was drawn by the sensational tingling in between the eyebrows as I mediated. Sometimes, the third eye opens without being in a meditative state or mind, and in this case, we look for answers that we can see, hear, or touch in our physical world. I will elaborate more on the subject in the upcoming chapters dedicated to our senses and their role while meditating. Sometimes, I feel the need to sit down and meditate – not in a quiet place but wherever I can. It could be in a loud place, or even outside while going for a walk.

Throughout the years, I read many books on spirituality and practiced daily meditation – sometimes a few times a day. I became an intense reader of spiritual books along with my school load. I learned to practice spirituality during intense moments, low moods, or on happy and less intense days. Everything varied form logistics, mood, health, determination, willingness and diet. While I lived with people in the house, I had a hard time practicing inside, thus I'd go to the beach –set a beach mat and get into guided meditation. Later on, I created my own altar, or a sanctuary where I did spiritual work and practiced healing, from physical ailments to emotional distress. Living with other people makes things harder, since they can be very dogmatic at times and they release their energy which can interact and can be intertwined with your energy, but I learned how to separate my physical body and create my physical space.

When people are "enlightened," meaning they want to explore spirituality, they usually search for answers, people to guide them. I realize that spirituality is very individual, and as such, it must be explained from a personal or individual perspective. It must be customized to one's desire, and aspirations to discover eternity. Simply put, my perception of the Universal Consciousness might be different from the person next to me. Note that the discussion about self, dates thousands of years yet the discussion especially in the last 2,500 years has been divided into two schools of thoughts spirituality or mysticism and science or rationalism. Yet both roads lead to the same conclusion, that of oneness, the unity between matter and energy, body and soul. Spirituality is the most tolerant of all "creeds and beliefs" and as such it allows you to have an opened mind while practicing. Spirituality is inclusive, positive, flexible, loving and forgiving. Spirituality can be practiced based on one belief or a combination of various. That is why I said it's inclusive. However, spirituality works best with Eastern beliefs such as Hinduism, or Buddhism and I will explain why.

Hinduism or Buddhism are not hierarchically organized religions like the Christian denominations are. Hinduism is more cosmic and universal than dogma based on one Deity, and as such it gives the believer more options. There is nothing wrong with Christian beliefs. Jesus Christ is a healer of body and soul as well as insightful, but when it comes to practicing meditation and other forms, a universal practice of spirituality is more open and receptive.

To sum up the argument, in this book the reader will find a deep understanding of inner self. The reader will be able to separate the socio-identity from spiritual identity. The book opens the door for many interpretations of inner self, mind, subtle body, physical body etc. I start the book with our realization of self-identity, then I offer a few philosophical arguments about the subtle body. After, I form a solid interpretation of inner-self, I look at the relationship between etheric body and quantum physics, reiterating a few theoretical points of view about energy, light, and the electromagnetic waves. In the chapters that follow, I offer an in-depth view of the mind, the conscious mind, our conscience and consciousness. I explain what relationship of all of the above concepts with the inner-self and the Universal Mind. When reading the book –as I mentioned above - one has to keep in mind that we all see the world in our own way, and we form our own perceptions. What appeals to me might not appeal to someone else and *nothing is absolute, not even the scientific explanation of the Big Bang despite the endless calculations. The world is a relative perception that we have about it and about the way it is run. But we have to keep in mind that the universe is a perfect creation whether you believe in God or not. We have no control over it. We surrender to its omnipotence. Even if one calculation was to be wrong, even if the number of degrees of power in the famous Einstein's formula $E = MC^2$, would slightly change, we would cease existing. And this brings me to a point that I should make at the end of the book, but I'd much rather make it now. Whosoever is in control of the universe, has done a damn good job. The universe is in perfection and in harmony. We have to align ourselves with this perfection. Let's take a look at our planet earth that we have claimed ownership over. Well, we can all agree that it is a total disaster and in deep state of chaos. This point illustrates how incapable humans are to control their destiny when guided by greed instead of principles and laws, including the laws of nature. In this book, I show a few ways of how to align with the divine through meditation and evade the chaos. My aim for this book is to offer an alternative understanding of self that thus far many disciplines have failed to address. Philosophers, scientists, clergy have always been perplexed by* *these paradoxes: 1. The physical body decays, but the subtle body does not. 2. The matter behaves like energy and the energy behaves like matter and both are components of the human body and therefore life. 3. The human mind connects with the Universal Mind, yet the human mind is product of a physical process whereas the Universal Mind is the totality of all actions - unseen. An explanation that requires a cohesive interpretation about self by bringing various points of views in the debate. The book is a way to find out what our inner-self happiness is within if considered without preconceptions. The sort of happiness that derives from feeling spiritually fulfilled. In the last part of the book, and for those who want a more hands-on approach. I focus on meditation practices, rituals, and since physics is such a wide discipline, I will also focus on the astral bodies and the connection between people and planets, how planets change our mood, behavior and how to make the most of your meditation practices. I will go in depth of the various meditation practices, starting with simple relaxation methods, to more sophisticated ways of achieving mindfulness and spiritual awareness.*

The core of this book

This book offers a deep understanding of self in many dimensions. I wrote this book to fulfil a dream of sharing my experiences with the reader and revealing what a wonderful journey spirituality is. It has helped me know myself better, discover myself and understand my purpose in this world. The book is written for those who would like a new perspective in life. That perspective is a different realm and the book will open a window where you will no longer need to wait for answers. They are right there in front of you. The book is written as comparative studies in philosophy and spirituality. The book offers research findings from Quantum Physics, spirituality, philosophy and cognitive psychology. A few years ago, I'd purchased many books by a divinity doctor who has written my practical books in meditation, chakras, ascension etc. She has an understanding of spirituality quite like mine and in one of her books she has explained her point of view in a graph which I found very interesting.[6] Yet while she labels the bodies and their properties, she lacked a binding element that keeps together the physical and the spiritual body, and that is the "relationship" of energy and matter. I have expanded my framework by adding more contextual meaning, such as the

---

[6] Shumsky, S. G. (1996). *Divine Revelation*. New York: Fireside.

relationship between the energy, the matter and the chakras which are wheels of energy. This way, the reader creates a mental image of my theory about self in spirituality. The table serves as a framework for this book and it touches upon the main points that I will be discussing.

Framework

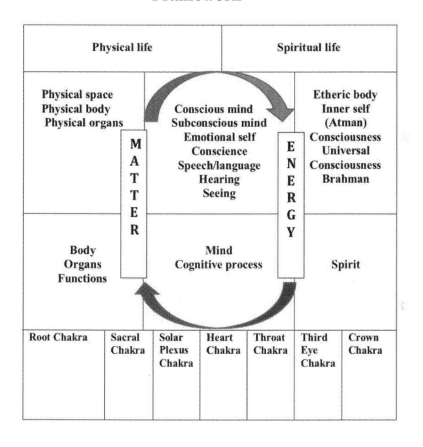

A SOCIAL VIEW OF SELF

This chapter deals with issues of group and social identity. Spiritual identity is not the first identity that we discover about ourselves. As a matter of fact, some people live all their lives without the slightest concern about spiritual identity. In social sciences, identity is the manifestation of the relationship between the individual and the society. For that matter, identity does not happen in a vacuum; but it is closely related to the social structure. Scholars have grappled with the idea of discovering how identity works, nevertheless, identity is a very malleable construct with many intrinsic and extrinsic factors that come into play. Identity is a social construct, and it occurs once the individual becomes aware of the relationship of self with the environment (external setting) and the others. There are many interpretations, and most of them are from a constructivist approach. Starting with the theory of self, Burke and Stets argue that self is rather an organized set of processes within us that accomplishes these outcomes (reflexive behaviors toward self and others). The question here is how we understand the self in social structures and how social structure shapes our understanding of self. The overall identity theories explain self in the dichotomy with other, in other words, what makes the self, different from the other. More importantly, identity, as we have mentioned, is a multi-dimensional concept. Cognitive process, language, and the social environment are crucial in understanding how identity works as an inner process. Identity starts in the "singular" form as *self* and is only recognized in a social structure where the interaction through language (internally and externally) becomes possible.

Another important concept in understanding identity is the "role." Simply put, we each have roles we assume in our society and daily lives, and these roles attribute many identities to us. By expanding the roles that people have in their lives, self also changes and becomes multidimensional. This characterization of self is limited only by the external environment and the "role" is forged identity. Self – as a construct- remains in a small realm otherwise known as the small selves as James (1980) characterizes it. Burke and Stets argue that self is the one's awareness about his or her identity. In my understanding, identity evolves around four major

ideas, a. the individual and their cognitive abilities-
b. social structure c. interactions with the
environment and d. interactions with other
individuals. Once many selves become aware of what
they have in common, then we have a collective
identity. The collective identity mirrors the group
convictions and beliefs. In addition, collective
identity must rely on the group's membership – an
association that the individual shares with others. In
such case, the membership can be an organization
or a political party. Or even belong to a particular
community. In this case, we have a new concept
called social identity. [7]

A critical view of social identity

From a social science perspective, social identity - as
a group identity - relies on shared interest to
determine adherence to a particular group. There
have been many interpretations of social identity but
for scholars of social sciences, the relationship
between the individual, the group and the social
structure is a loose concept – meaning, the
individual can construct and choose social
identities. For example, Oakes argued that shared
interests of a group can be a wide range and

---

[7]Burke, P. J., & Stets, J. E. (2009). *Identity theory*. Oxford: Oxford
University Press.

categorized based on two processes- *accessibility* and *fit.*[8] In other words, social identities are available to people, and people have the choice to pick their own social identity. Oakes identifies accessibility as the availability of a specific group to its members so that they can identify themselves with the group. The idea is that *accessibility* must be available for individuals at all times. Such category includes race and gender- for example. On the other hand, *fit* is mostly described as the appropriate social identity where people can match their interest in the group's motto, philosophy, beliefs, etc. Deducing from this theory, Oakes' *fit* is the alternative to accessibility in case persons do not identify with any of the accessibility options. These technical definitions of social identity can be ambiguous, and while coupled with political science concepts and ideologies they become more complex.

The fit as a process allows people to choose their social identities such as religion, or political affiliation. [9] Hogs and Abrahams argue that social identity is inherent to a person's identification with a social group. "A social group is a set of individuals who share the view that they are members of the

---

[8] Ibid
[9] Ibid

same social category." [10] Through a social comparison and categorization process, persons who are similar to self, are categorized with the self and are labeled in the group. More importantly, "Persons who differ from the self are categorized as the out-group." [11] This interpretation stems in the social identity proponents Henri Tajfel and John Turner whose view on identity is a constructivist method of analysis, categorized in three cognitive processes: a. Social categorization, known as the process of deciding which group you or another person or person belongs to. b. Social identification, the processes by which you identify with an in-group more explicitly and, c. Social comparison, when a person's self-concept or social concept becomes closely interconnected with perceptions of group membership. Social identity defines the individual by the common interest and perspectives. For example, similarities within the group is a common or shared interest, meaning, people in the group think alike and act alike and that distinguishes them from the other groups. [12] In this case, common interest is a behavioral construct more so than

---

[10] Cited in Identity Theory p. 118
[11] Burke, P. J., & Stets, J. E. (2009). *Identity theory*. Oxford: Oxford University Press
[12] Ibid

ethnic or racial homogeneity. For example, Stets and Burke argue that social identity gives members of a group a sense of belonging and with the sense of belonging, a sense of self-esteem. [13]

Self-esteem, as Cast and Burke argue, is often rooted in the reflected appraisals process, in other words to feel valued and appreciated.[14] Moreover, Cast and Burke claim that high levels of self-worth, give people the existential security, which provides value and meaning to their lives. *When one is a member of a group is similar to others in thought and action, one will receive recognition, approval, and acceptance from other group members, thus verifying their social identity as a group member; and in turn, they will experience positive feelings.*[15]

Individualism vs collectivism

Understanding self is knowing first what you are not, then what you are about. In a broader definition of self, we tend to frame our persona to the groups that we belong. However, the inner-self – the identity that makes us unique often lacks from our understanding. Social sciences are not concerned with the inner self in a broader sense. They study

---

[13] Ibid
[14] Ibid
[15] Ibid

groups, and behavior patterns in many cultures and societies. This approach perhaps answers a few dilemmas people have with identity crisis and soul searching. Ideas that people have been trained to frame themselves into.

We often hear people say:

I have not found myself in a certain job, in my marriage, or various settings. Or have you heard someone say? I have a feeling that I don't belong here or there. Strangely enough, people are always searching for the sense of belonging to the group or community and perhaps they search all their lives in vain. Later, we will also discuss that the world can also be an illusion.

Inner self is pure essence. It gives people the freedom to understand their existence to find their true purpose in life through mindfulness. This approach defeats the purpose of societal "constructs" and believes we take for granted.

While we belong to a group that we identify with, and thanks to our conscience, we develop sets of values and morals. Most of the cultures and societies teach morals and values since an early age. Youngsters are taught to embrace these values without questioning that these principles are group norms and not individual freedoms. Yet these norms keep us grounded. Societies take the idea of shared values to

30

the next dimension. Nowadays employees are taught to think collectively, and not individually. Many companies, in the private and the public sector train employees to embrace common shared values. By doing so they replace individualism with collectivism. I resented that type of training.

Whether we agree or not, this type of societal programming is against individualism, the core of human nature, and it demeans who people truly are inside - it diverts us from feeling free. It places constraints to our existence. It shows lack of faith in humans as individuals. It erases the self-consciousness to replace it with collective thought and pragmatic ideas to serve larger interests by underestimating self-interest. When we think collectively, we no longer belong to ourselves, we belong to the group. If we are stripped of individualism, we can no longer be our individual selves.

People in general have that awareness in them and know right from the wrong. We do what is good because good is the path to happiness – harming might please someone's ego, but it cannot escape their conscience and deep down the harmful person knows their deeds. Those who harm have deeper problems, not only societal, but also individual and psychological. The history of humanity has shown that this is perhaps one of the reasons why societies are so adamant about instilling values. People must follow the societal rules, but without undermining the uniqueness inside every one of us.

On a larger spectrum, societies and elites need uniformity in many shapes and forms to control the masses through political systems, religion, economic systems, cultures and so forth. While the opposite would be an anarchist idea, it is important that we see individuals in their mere definition of the "individual." Adhering to identities mentioned above transplants the individual from their true self into the norm, the collective. I am not promulgating a libertarian idea either, yet I am emphasizing that promoting collectivism at the expense of individualism, it is perhaps a disservice to our true self and self-expression.

Many times, students of mine have told me through the years: Well I don't want to be like everybody else. I took these comments as spiteful disrespectful. We are trained to think collectively. Doesn't the society do the same to individuals who think differently? They are ostracized, not hired, ridiculed for being unique and for having a good understanding of their inner self. Many of these brave souls have suffered the consequences and live in the shadows of the society. Nevertheless, I must say that they have found their core, their essence and know who they truly are. These individuals find no significant meaning in groups, or organized religion, or even a highly ordered society, with rules and restrictions.

For example, Friday, Saturday or Sunday depending on religion, people go to the cult to hear the priest, the pastor or the imam say, "You must not sin. You must not do certain things" over and over. The rituals are repetitive, and not to scorn any religion, but while something becomes repetitive and redundant, it loses its meaning. People go to the cult buildings no longer to get in touch with the higher power –but to fulfill the obligation and to feel that they are doing the right thing without benefiting one bit from the Divine Spirit, and the higher power. Many followers are not taught to get in touch with the higher power, but to have an intermediary pray for them. The moment people give away their God given power to get in touch with the divine, they have lost that gift, and they gladly give it away. On the other hand, meditation will restore this gift and skill right back. Have we asked ourselves, why do we have institutionalized religion? Because we (some not all) are underestimating ourselves, our power, who we are, the power within us, and we have become prisoners of the cult while we are overlooking God right in front of us.

In conclusion, we are all social products and we exist and function well in a society. While we are part of the group or the community, or even a private enterprise, we still need to preserve that identity. Collectivist thought must be balanced and must not replace the individualism and the individual freedoms. Many societies have a collectivist organization, yet these societies do not consider the person to be an individual, but a number.

A THEORETICAL VIEW OF THE INNER-SELF

Thus far I have explained how the concept of self, to social sciences is a collective concept. Many scholars see self and identity from a constructivist point of view. Simply put, they build the concept based on studied societal behaviors and patterns. Keeping in mind that sociology sees the individual as part of the group, the dichotomy self and other only explains the uniqueness from a collective point of view. Social sciences are not concerned with a metaphysical interpretation of self. Two of the oldest religions that are in my opinion philosophies – Hinduism and Buddhism have been long concerned with the understanding of inner self, true self, the soul and the Universal Soul. In the old Sanskrit scriptures, it is called Atman (Individual soul) and Brahman (Universal soul). Later on, the Greek philosophers such as Socrates and Plato, have tried to separate the physical self from the subtle body that they call soul and argued that that once we are able to liberate ourselves form our attachment with the physical world, we can experience eternal life and bliss. Thus, the universe presents us a world which is physical and concrete at the same time the universe mirrors a subtle world that is the replica of our reality. This subtle world we call spirituality starts with the smallest particles of cosmic energy to powerful binding electromagnetic waves that keeps us all intact.

Thus, concerned with understanding self-rose as people needed to understand how we relate to the Universal Consciousness, the higher power. There are many interpretations of the inner self in religion, and some of them break the barriers of institutionalized religion. In general, institutional religion is meant to create cultist interpretation of the spiritual world to control their followers without giving them a chance to think for themselves.

The old religions were concerned with understanding the relation between self and whole or totality such as the connection between man and God. In the mere sense of this connection we always conceptualize it in a physical world. Thus, we follow the cult in the physical world, study the scriptures in the physical world yet we are not taught or trained how to connect in the spiritual world. The institutionalized religion and many dogmas out there are only concerned with morals and values and not the realization of self-consciousness and the pathway to God.

Buddhism

Buddhist Criticism of the Metaphysical Self

Buddha – a male born approximately 2,500 years ago to a very affluent family, became a symbol of spirituality. In his place of birth Nepal, among other teachings Buddha showed to the devotees, that the physical body brings people sufferings, is not permanent, and one must unify with the divine to achieve the state of Nirvana.[16] These were the basic ideas of self in the Buddhist philosophy. Buddhism unlike Hinduism has a different perception of self. In principle, when it comes to realizing self, humans must practice detachment from the existence which according to the Buddhist philosophy means suffering – both mentally and physically. Thus, as the individuals understand that their existence is characterized by impermanence, suffering, and [non-self], then we learn to detach ourselves from our existence and live in nonphysical, timeless, and spaceless form. This type of existence is otherwise called Nirvana. It is a bit hard to grasp. In simpler

---

[16] Rahula, W., & Demiéville, P. (1962). *What Buddha Taught*. New York: Grove Press.

terms: Buddhism denies people of recognizing their "inner self" because it brings suffering. I assume it is the ego that drives the suffering. In the case of humans, Buddhist philosophy says, "No factor of human existence is everlasting; it has its beginning and end." They necessarily produce Samsāra,[17] the Empirical world and different kinds of duhkha (suffering).[18]

Anattā in the early teachings was otherwise known as the concept of non-self.[19] It is one of the three principles mentioned above and what distinguishes Buddhism from Hinduism. Thus, Buddhism sees the human being as not having *a soul* since humans submitting, to their physical existence, go through suffering and impermanence. In the early teaching of Buddhism, the doctrine identified self as the soul and as such – it argued that it did not exist. Fundamentally, this view of Anatta,[20] was later rebuked by modern Buddhist philosophers.

---

[17] ULE, A. (2016). The concept of self in Buddhism and Brahmanism: Some remarks. *Asian Studies.*
[18] ULE, A. (2016). The concept of self in Buddhism and Brahmanism: Some remarks. *Asian Studies.*
[19] Richard Gombrich (2006). *Theravada Buddhism.* Routledge. p. 47
[20] Christmas Humphreys (2012). *Exploring Buddhism.* Routledge. pp. 42–43

Hinduism: True self or Atman

Understanding Atman is as easy as understanding self. Vedanta says, if you understand yourself – then your suffering will come to an end. What is the self in Vedanta? Is the person I think I am, or something I don't know, or beyond that. You are the real self, your consciousness. In the light sleep, in the deep state. – These conversational instances are recorded from a speech Swami Sarvapryiananda gave a few years ago. [21]

In Hinduism the inner self relates to Atman[22] and Atman is part of Brahman – the greater God. Thus, Atman is inside of us, and by its presence, we realize that Brahman (God) is inside of us, in a part and a whole relationship. It is very important to understand a few concepts: We are human beings – and as such we experience ourselves through our physical bodies. Hence by being aware, it enables us to have an earthly experience, to sense the world. Our functioning capabilities such as the sensory system to hear, listen, touch, smell, taste, and lastly [the mind] help us become aware of our inner self and the world around us.

---

[21] Vedantany1894. (2018, December 19). || the atman || by Swami Sarvapriyananda. Retrieved April 15, 2021.
[22] David Lorenzen (2004), The Hindu World (Editors: Sushil Mittal and Gene Thursby), Routledge, pages 208-209

The mind is not part of the sensory system, but it is the process responsible for our cognitive abilities and the thought process. It is a tool that enables us to experience our inner self besides the world. It is exactly the mind that enables us to make the connection between the subtle body and the physical body, the world and the Universal Consciousness. The inner self according to Hinduism goes beyond death. The physical body and the mind are changeable, whereas Atman –God in us is unchangeable.

How do we understand the eternity of Atman?

Atman is present before our birth and also after our death, says Vedanta in Hinduism, the understanding of true self or inner self relies on the separation of the physical body, subtle body and causal body. Hence, according to various schools, we experience our physical bodies from birth to death. That is called life. During this time, physical bodies go through transformation, and slow degradation as we approach the natural process of death. And death means the end of a cycle and beginning of a new one. The end of the cycle can also be deemed as liberation from constraints and suffering that keeps us bonded to the physical life.

Apart from our physical body, we also have our etheric body also known as the subtle body. Many are not able to experience the subtle body because – it is in our human nature to only experience the physical and materialistic world. Our mind, thoughts, and feelings are part of the subtle body. Thus, in Hinduism, the subtle body is none other than Atman – the soul. [23]Note that to the etheric body we add the consciousness to make it complete, pure, and essential.

On the other hand, the causal body is a subtle body that carries out the process of reincarnation from one individual to the next. Thus, the causal body is the soul in between two lives. A dying life, and a birthing life. Interestingly enough, that transitional period prepares the soul for a new journey; a new life.

The principle of casual body is associated with one or more *koshas*. The koshas are the layers of awareness that veil *Atman*, or true Self. [24]*"The causal body contains the anandamaya kosha (bliss), where the yogi experiences calmness, peace and joy. Some traditions refer to this layer as the true Self, while others believe this kosha simply opens the door to the true Self.*[25]

---

[23] Deussen, Paul and Geden, A. S. The Philosophy of the Upanishads. Cosimo Classics (June 1, 2010). P. 86.

[24] Ibid

[25] What is the causal body? - definition from Yogapedia. (n.d.). Retrieved April 15, 2021, from https://www.yogapedia.com/definition/5802/causal-body

In essence, Atman cannot be separated nor understood without the presence of Brahman (Almighty God).

In Hinduism, Atman is never born, nor it ever dies. Atman is present before birth and after birth. While Brahman is the totality where Atman returns in union after death. Brahman is the source, the truth, and absolute reality.[26] The relationship between Atman and Brahman is a relationship between self and totality. Brahman is God without attributes, limits, or characteristics which Hindus call Nirguna Brahman. Brahman and Atman can be experienced at a state of higher consciousness. [27]

Ancient Greece: Sages and philosophers
Socrates

Greek philosophy is the foundation of today's modern society and modern political systems. Yet Greek mysticism and rationalism were once inseparable. Hence, the Greeks' understanding of the world was based on macrocosmic interpretations. In discovering the world, Greeks started their journeys with the study of the universe;

---

[26] Achuthananda, S. (2013). *Many, many, many Gods of Hinduism: Turning believers into non-believers and non-believers into believers.* Charleston, SC: CreateSpace Independent Publishing Platform.
[27] Ibid

all things have an essence they called phisis.[28] While they defined phisis, Greeks moved to matter and discovered that matter was made by small particles they called atoms. Cited in Karpouzos:

The word atom is derived from the ancient Greek word atomos, which means "uncuttable." This ancient idea was based in philosophical reasoning rather than scientific reasoning, and modern atomic theory is not based on these old concepts. That said, the word "atom" itself was used throughout the ages by thinkers who suspected that matter was ultimately granular in nature.[29]

Yet, in the early math and scientific calculations, Greeks were fascinated with astrophysics. For the early Greek sages, the universe was created by an omnipotent force. Their cosmic reasoning derived from calculations and theories that were a mix between rationalism and mysticism. For Anaxagoras, Archimedes, Thales and many more Greek mathematicians, cosmos was the perfect order of things. Alexis Karpouzos, a Greek philosopher who tries to build the bridge between science and spirituality writes:

---

[28] Karpouzos, Alexis. *Universal Consciousness: The Bridges between Science and Spirituality.* Athens, Greece: Think.Lab, 2020.

[29] Pullman, Bernard (1998). *The Atom in the History of Human Thought.* Oxford, England: Oxford University Press. pp. 31–33.

"Thales[30] declared all things to be full of gods and Anaximander[31] saw the universe as a kind of organism which was supported by 'pneuma,' the cosmic breath, in the same way as the human body is supported by air. Pythagora's Cosmo-theory supported that spirit is the matter of the world and it is subject to a mental set that expresses the Universal Divinity." [32]

Later on, Socrates argued that it's the voice of God that works inside us. He assumed that the conscious mind was in fact God speaking directly and that voice is an echo of the voice that governs the Universe and defines the operation of everything in the world. Thus, Socrates was one of the first Western philosophers to bridge that connection between inner self and God. He contrasts yet connects the "cosmic creation" and the "human fate." [33] His understanding of self does not differ much from the Eastern perception of inner self, or subtle self. Socrates distinguished between the

---

[30] **Thales of Miletus**, (born *c.* 624–620 bce—died *c.* 548–545 bce), philosopher renowned as one of the legendary Seven Wise Men, or Sophoi, of antiquity. (Britannica)

[31] Anaximander (/æˌnæksɪˈmændər/; Greek: Ἀναξίμανδρος Anaximandros; c. 610 – c. 546 BC) was a pre-Socratic Greek philosopher who lived in Miletus, a city of Ionia (in modern-day Turkey). He belonged to the Milesian school and learned the teachings of his master Thales. He succeeded Thales and became the second master of that school where he counted Anaximenes and, arguably, Pythagoras amongst his pupils. (Wiki)

[32] Karpouzos, Alexis. *Universal Consciousness: The Bridges between Science and Spirituality.* Athens, Greece: Think.Lab, 2020.

[33] Ibid

mortal body or the physical body and subtle body or the immortal body. The duality in which Socrates lays his foundation of the understanding of self is the juxtaposing of two concepts, the changeable nature of the physical world and immortal or nature of our eternity, known as "unchangeable being." [34]

As in the Eastern philosophy, Socrates believed that the physical world is a changeable world and it enables us to live an experience though our senses. Thus, our physical existence is subjectified through what we can see, hear, taste, smell, and feel.

In essence all aspects of our physical world are continually transforming, changing its form, disappearing and reappearing again. The eternal, on the other hand, is unchangeable, and that includes the truth, the universe, even philosophical concepts such as beauty, and goodness. [35]

Just as we saw in the Hinduist belief of self, for Socrates our bodies were imperfect, they went through transformation from birth to death and they belong to the physical world. Whereas the soul – Socrates said, belonged to the "ideal realm." The soul survives the physical body after death. Socrates was

---

[34] Plato, Benjamin Jowett, and Irwin Edman. *The Works of Plato*. New York: Simon and Schuster, 1928.

[35] Chaffee, J. (2015). Philosopher's way: Thinking critically about profound ideas. In *Philosopher's Way: Thinking Critically about Profound Ideas* (pp. 92-104). Pearson.

able to establish a relationship between the soul and the body, though he recognized that these two are two different entities. Socrates had a dualist interpretation of the body and soul and he argued that the mind is the tool that makes the connection between these two.[36] Socrates thought that in the physical world, our existence as humans, has the power to drag the soul along because we – as physical beings- are imperfect in our nature. Thus, the soul is none other but a weaker entity that might wander around "confused." In that sense, Socrates mentions that the only way the soul will find its way to a perfect and unchangeable world is when the power of the reason overcomes all the obstacles and takes over. The reason –according to Socrates – can return the soul to the unchangeable. Thus, in Socrates' explanation of two realms and their relation, we see a few parallelisms with the Eastern understanding of self. However, In Socrates' perception the reasoning – which will be self-discipline would cause the liberation from the evils of the physical body to the unchangeable realm. Later on, in this book we will discuss the power of mind and what the impact has on spirituality.

---

[36] Chaffee, J. (2015). Philosopher's way: Thinking critically about profound ideas. In *Philosopher's Way: Thinking Critically about Profound Ideas* (pp. 92-104). Pearson.

Descartes (Cogito Ergo Sum) I think therefore I am

René Descartes was a French philosopher. He lived in the seventeenth century. Descartes is mostly known for his famous quote in Latin, "Cogito ergo sum - I think therefore I am. (Je pense donc je suis). There are various interpretations from a philosophical point of view, the most generic one is that Descartes, implies is that the only thing that I am not doubtful of, is my existence as we live in a world where many things are an illusion. This interpretation has been criticized by many. Descartes literally places thought before existence. Yet, Descartes perhaps was not wrong. Philosophers and critics refer to his theory as dualism. Note here, that duality is not the relationship between man and God, but the duality between mind and body.

> In dualism, 'mind' is contrasted with 'body,' but at different times, different aspects of the mind have been the center of attention. In the classical and medieval periods, it was the intellect that was thought to be most obviously resistant to a materialistic account: from Descartes on, the main stumbling block to materialist monism was supposed to be 'consciousness,' of which phenomenal

consciousness or sensation came to be considered as the paradigm instance.[37]
With the birth of modern science, the dualism between energy and matter became more prominent as the inquisition worked to stop progress and punished many voices and scientists who tried to have a scientific posture about the universe. In the seventeenth century René Descartes became known for his fundamental division between two realms: Mind (res cogitans), and matter (res extensa).[38] What philosophical approaches lacked in the interpretation of Descartes is the affirmation he'd made about his spirit and his mind as the catalyst in understanding one's existence. Also, many philosophers have a materialist point of view and seldom refer to energy as "spirit." For the materialists, mind and brain are not separated, mental states and physical states are the same.[39]

> *Descartes argues that the clarity and distinctness rule, derived from the Cogito, can justify our beliefs about the external world. But what verifies the clarity-and-distinctness rule? God's existence, Descartes argues. By*

---

[37] Robinson, H. (2020, September 11). Dualism. Retrieved April 18, 2021,
[38] Karpouzos, Alexis. *Universal Consciousness: The Bridges between Science and Spirituality*. Athens, Greece: Think.Lab, 2020.
[39] Robinson, H. (2020, September 11). Dualism. Retrieved April 18, 2021,

reflecting on his idea of God, he argues that God exists. Descartes then argues that a truthful, good God would not allow us to be deceived when we understand objects clearly and distinctly, *and so God would not allow us to routinely have false beliefs.*[40]

Thus, in this philosophical interpretation, the scholars entirely miss Descartes' point. Descartes is not questioning the existence of God and whether he lives in a deceiving world. Descartes is in fact trying to build that bridge between inner self and Universal Consciousness, as mirrored in his work "meditations." He is not arguing about false beliefs, and illusions, but he is pondering an unseen world that he can experience through trans-meditation.

Like so, earlier from a Hinduism explanation offered by Swami Sarvapryiananda,[41] the mind is part of our subtle body. The mind does not dissolve, nor does it die at the moment of death. But we do. When we leave this world – our physical body dies. The mind still remains a mystery. The mind is essence and thoughts are energy. The interpretations in my view are pure and simple. Descartes here has –

---

[40] 1000-Word Philosophy: An Introductory Anthology. (2021, February 16). "I think, therefore I am": Descartes on the foundations of knowledge. Retrieved April 18, 2021.

[41] Vedantany1894. (2018, December 19). ‖ the atman ‖ by Swami Sarvapriyananda. Retrieved April 15, 2021.

distinguished between the physical body and the inner self. I exist, he says, and I know this, because I think. He places emphasis on the thinking process which we will explore further in the chapter where I talk about the mind theory. Descartes attributes his awareness and consciousness to the mind and the thought process, with which we are able to go beyond our physical existence. [42]

The difference between the Eastern and Western schools of thought

There is no substantial schism between the Eastern and the Western school of thoughts and their interpretation other subjects of the truth about the universe, our existence, and our relation to the Universal Mind. They seem to agree in the relationship between energy and matter, idealism and materialism or rationalism and spiritualism. Some of the substantial differences have to do with the interpretation of the inner self and the Universal Mind. The key point in these schools of thought is the recognition of the inner self as part of the Universal Consciousness and the human mind as the medium that facilitates such connection. The

---

[42] Chaffee, J. (2015). Philosopher's way: Thinking critically about profound ideas. In *Philosopher's Way: Thinking Critically about Profound Ideas* (pp. 92-104). Pearson.

Western school of thought often confuses the mind with the inner self, while it agrees with the concept of self-consciousness as a product of human brain integrated through the mind. Thus, what we need to understand is that:

The mind is related to the brain- the physical organ and when the physical body dies, the self-consciousness returns to the Universal Consciousness; the storage of greater data. Self-consciousness is only made possible by the mind as a process. The thought is the catalyst for people to develop a self-consciousness. If the thought process is interrupted or impaired, so will the consciousness.

The philosophy of Descartes was not only important for the development of classical physics, but also had a tremendous influence on the general Western way of thinking up to the present day. Descartes' famous sentence 'Cogito ergo sum'-'I think, therefore I exist'-has led Western man to equate his identity with his mind, instead of with his whole organism. As a consequence of the Cartesian division, most individuals are aware of themselves as isolated egos existing inside their bodies. The mind has been separated from the body and given the futile task of controlling it, thus causing an apparent conflict between the conscious will and the involuntary instincts.[43]

Thus, in the above explanation I underlined that consciousness is a part of the inner self. Inner self is what connects us to the divine and revels our true existence: Inner self in the Eastern school of thought is nonetheless then the subtle self of our existence, the Atman (God in us). To westerners it is a bit difficult to understand and grasp the Eastern philosophy, yet in previous decades there's been a shift. More people in the West are showing tremendous interest in the Eastern thought, religion and practices.

---

[43] Segrest, D., & *, N. (2020, September). Archetypes and symbols. Retrieved April 18, 2021.

THE SIPIRTUAL SELF

Self is such a profound concept that is not limited to the physical self as most of us understand it. In this book, I analyze self from many perspectives and on many levels and in this chapter, I consider self in a different realm therefore, bringing forward the subtle self, or inner self. I focus my efforts in explaining the inner self, how we understand and interpret it. Then I explain the etheric body and the difference between these two. Remaining on the subject of self, I conclude that, the experiences we have once we enter spirituality are not the unseen, but the reality in which we exist and is accessible to everyone.

Understanding of inner self

"O Inner-Self, may your voice be clear and loud: may you become the oars of the boat of our life, guide us. Guide us to the safe shores of the boat of our life; guide to the safe shores through the stormy sea of life, avoiding all calamities." Vedas (2.42.1)

In Buddhism, the interpretation of self is very rigid, at least in the old scriptures. Paradoxically, the concept of inner self exists, but it is called the no-self, or no-soul, and this way, union with Buddha liberates the soul from the suffering.[44] In Hinduism, the concept of inner self in not only sophisticated, but it is a major part of the creed. Self cannot be separated from the larger soul: Atman cannot be separate from Brahman – they are in an everlasting relationship. [45]Yet, to appreciate self, one must be able to separate it from the physical body. Inner self is pure essence; it is the entity known as our subtle body. We cannot see our inner self unless we step back and realize its purity. Inner self, or as the Hinduism refers to it as Atman, is pure self-realization, in coherence and a vital part of Brahman – (God).

Seldom people experience awareness of their subtle body – the inner self or the soul, because we are taught to unite our soul with the Deity (God). How can we be in union with God, when we are not aware of our inner self, and our soul. How are we uniting with God if we have never experienced our inner self in its own totality?

---

[44] See "What Buddha taught" Rahula, W., & Demiéville, P. (1962). *What Buddha Taught*. New York: Grove Press.
[45] Vedantany1894. (2018, December 19). || the atman || by Swami Sarvapriyananda. Retrieved April 15, 2021.

The union with God it is a sublime moment, and just because we say it, or read it in the scriptures, does not mean that we are doing it.

Hinduism has one of the most explicit approaches when it comes to describing the inner self. In a real world, inner self is not the voice that hums inside the heads, that is the voice of conscience and is deeply influenced by the environment, the experiences in the physical life interactions with others, our cultures, prejudice upbringing etc. Inner self is other than conscience and is apart from the physical body. To clarify the misconceptions of self and many bodies[46] that (we) as people carry in our existence, it is important to recognize – what inner self is like, and what happens to it when the body dies. As humans, we live in a physical world. The existence in this world is defined by the purpose each one has. Each and every one strives to find that purpose in life. People are born, live and die in the physical world. The body disintegrates after death and people accept the fact that the physical existence is temporary. While life on earth can be as complex as mysterious, people are equipped with an innate mechanism that is universal to all to better understand this world. This mechanism enables the thinking and it pushes the thought process beyond what is seen, heard, and

---

[46] Referring to physical body, subtle body, inner-self, soul, etheric body.

perceived. Humans have always looked for answers if eternity exists, or there is after-life. Descartes said, I think, therefore I am. The mind is what deems us people as intelligent beings, able to ponder about other realms and the metaphysical world. Thus, the brain, and the mind is what generates the thought process. However, the brain is a physical organ and the mind is subtle, and so are the thoughts. The mind cannot be seen, touched or felt, nor can people read each other's minds using their physical powers. The mind acts as a mechanism which connects the physical body with the spiritual or subtle body and eventually the higher power. The mind is separate from the physical body, yet for as long as people are conscious about their existence, the mind functions within and is highly individualized. The mind is the gate, the conduit that opens up into a deeper state related to the existence – the inner self. This state can be achieved through meditation, and the meditative mind. The role of the mind in reaching out to the inner self is paramount –and I will discuss in the next chapter the theory of mind and argue that, impairment of the mind through disease, mental disability or addiction to substances, will prohibit the mind to reaching out to the inner self. Understanding inner self is not easy, for many reasons. First, it is understood that the inner self equals the conscious mind. Secondly, the mind is a very complex process – It is a computer program that gathers and processes information. When a person enters a meditation state, the mind takes the role of the guide. Life on earth is built a certain way that people cannot function without the power of their mind. And as such the mind is programmed to satisfy the immediate needs of each individual. Materialistic gains, financial gains, even in the early signs of society where men were hunters and gatherers, were achieved by the power of mind and reasoning. The mind facilitates that instinctive need for survival. Whereas, people strive for emotional stability and happiness because everyone must reach a degree of satisfaction that is pleasant and fulfilling. It is still the mind's job to find resources to satisfy the emotional needs. Hence, people seldom worry attaining spiritual bliss because it hardly materializes in financial gains – maybe for some, but not for all. On the other hand, some people have little, yet a rich spiritual life. There aren't many studies about inner self. Inner self is completely separated from what's "substantial" in the materialistic world. Ancient civilizations have dedicated a whole lot of effort getting in touch with the divine to remind us that the search for eternal bliss stems with holy scriptures of perhaps the oldest religions such as Hinduism. Thus far, self is highly personal and perceived as nuclear identity. No one can ever escape the inner self, no one can hide from the truth. Inner self is the inner side that everyone protects and keeps sacred. It is in human nature for people to shelter what's sacred and personal. People realize their own identity very early on, and once that identity is established, people become aware of the dichotomy of self and other. The inner self is not only the character, but the driving force that makes individuals unique. The inner self includes character, desire, dreams, failures, pain, happiness, vice, morals, values, perception and the past. The inner self can be strong, it is often shelled and protected with love, care and respect for ourselves. If left unprotected and non-sheltered, the inner self will be harmed, and in here I am not implying to hurt feelings and egos, but I am referring to events that will shape one's life and destiny. There are many people out there that are dear to us who are going through perpetual pain and suffering. These people have lost their soul, or their soul has been crushed in one way or another. Such as experiencing a difficult childhood, abusive relationship or marriage, substance abuse, PTSD and mental disease, and many other factors that would impact a person in a very profound way. They have lost their inner self and many, will never find it.

The inner self is considered to leave the physical self/body and it is believed to continue to live and return to the Universal Soul – as I have mentioned before. Thus, the subtle body – is the first exhibition of inner self; it is the outlet, the passage that connects with the Divine, higher power, God, Brahman or however people wish to refer to it.

Yet, when we ran our course through our physical existence, and after our departure from this world, what happens to our inner self, the subtle body, our thoughts? In fact, none of these die out or fades, because they do not exist in the physical world. They are energy, and they also leave the physical body to unite with the Universal Soul, Brahman, God. As naïve as it might sound, the returning of the soul back to the Universal Soul is as natural as entering a new life. To better understand the true self, let's do a little experiment:

> *"Walk to the mirror, as soon as you wake up take a look at what you see in the mirror." That is a reflection of you. Do the same thing, when you are dressed, adorned and decorated for a soirée or a party. That person is still you. Nothing has changed but the superficial side of it. Inside you exist, whether you're happy, or*

sad, whether you look fancy or natural. We cannot escape being who we are. Our true self. This allegory helps us understand our inner self.

The physical body's appearance isn't the true self, but what we perceive as our inner side reflected in the mirror is our true self. S. J. Chung (2012) argues that, human beings are composed of the body made of material matter, and the conscious mind.[47] Thus in this statement I agree that there is a physical self, made of matter, that we also refer as our physical body, our biological, genetic body. The physical body is responsible for creating a relationship with the physical world in a broader sense, the relationship with others, not only on the basis of kinship but also on a bases of larger social constructs. As Chung continues, "There seems to be another self in my manifest consciousness that is felt as 'I', and closely associated with my bodily functions and my brain activity, sensing the world around the body through sense organs. The latter self is associated with the physical body; I name it the 'physical self." However, Chung distinguishes a triad – the conscious mind, the physical body and the inner self. The last two

---

[47] Chung, Sung Jang. "The Science of Self, Mind and Body." *Open Journal of Philosophy* 02, no. 03 (2012): 171-78. doi:10.4236/ojpp.2012.23026.

are tightly related to the conscious mind, yet when the physical death occurs, the inner self– Chung suggest unites with the Universal Consciousness, or spiritual world. The conscious mind seems to be occupied by the physical self-and/or the inner self. [48]

The soul and the physics

Etheric body is known otherwise as the "soul." It has been subject of study for many philosophers, scientists who understood that people are more than a physical body. Etheric body is what we know as the energy field that we embody and carry throughout life. The aim of this book is not to complicate matters more for the reader but to break down the analysis in much more friendlier terms that's easy to understand what the etheric body is. Since energy is not discernable, it would be hard for the human eye to see the etheric body, but nevertheless, it exists. What people find hard to understand when making a relationship between spirituality or religion and science is the basic matter and energy relationship. People usually expect to see, witness, and experience miracles in the physical world knowing that God exists in the spiritual world. This is the biggest challenge that today's spirituality faces. People expect to see materialistic gains using energy and

---

[48] ibid

spirituality. The Law of Attraction has been misinterpreted and sold as a way to manipulate energy and gain financial means without establishing first an understanding how energy and matter work. We are made of matter, but we are also made of energy. How do we bring these two together to understand the essence and what keeps them bonded? Joseph Selbie in his book, *The Physics of God* summarizes that more open-minded scientists believe that the relationship between matter and energy does not explain all phenomena, (and as we see with Newton's observations), but it will leave room for more interpretations in the realm of spirituality and religion, claims—miracles, heavenly realms, life after death, personal transcendent experience, immortality of the soul, and God.[49] On the other hand, Einstein describes in his theory of relativity the relationship between matter and energy in the equation:

$E = mc^2$ *(energy equals mass times squared speed of light)*

In this equation, Albert Einstein develops an understanding of the theory of special relativity that expresses the fact that mass and energy are the same physical entity and can be changed into each

---

[49] Selbie, Joseph. The Physics of God (p. 168). Red Wheel Weiser. Kindle Edition.

other. In other words, the increased relativistic mass (m) of a body times the speed of light squared (c2) is equal to the kinetic energy (E) of that body.[50] Earlier, Sir Isaac Newton explained, the existence of the etheric body, or the soul as they relate to energy: Among others, Newton knew that we are made of energy, a vital part of our gross (physical body) which he explains "is the energy that binds the body together, the cells binding with one another." He then explains that there is a force that keeps these cells or particles together, the way they bond naturally with one another, and that is our etheric body – the energy field inside of us. I have found his writing about energy and matter in our physical existence very compelling and I have quoted it as follows:

(Adopted for the readers)

"And now we might add something concerning a certain most subtle spirit which pervades and lies hid in all gross bodies by the force and action of which spirit the particles of bodies mutually attract one another at near distances, and bind once they come in contact with one another; when electric bodies operate to greater distances, as well repelling as attracting the neighboring cells; and light is emitted,

---

[50] E = mc2. (n.d.). Retrieved April 18, 2021, from Britannica

reflected, refracted, inflected, and heats bodies. All sensation is excited, and the members of animal bodies move at the command of the will, namely by the vibrations of this spirit, mutually propagated along the solid firmaments of the nerves, from the outward organs of sense to the brain, and from the brain into the muscles." [51]

Thus, Newton recognized the facts of the etheric body. The etheric body is what we now refer to as the energy field inside of us that interacts with our physical body including our organs. It allows us to express ourselves in the physical world. At the same time, it integrates the physical body into earth's magnetic field. [52] Since the etheric body is pure light and energy, it serves as a blocker of all harmful things and agents coming to the body.

Close your eyes for one minute or two and know that there is light and a subtle body inside all of us. The etheric body is a protector, and an immune field not from a physical perspective or part of the immune system, but rather an affirmation of the psyche which tends to explain the basic survival questions: Am I safe? Am I comfortable? Am I free of pain? It represents the base survival level of our being. [53]

---

[51] Bailey, A. A. (1931). The soul and its mechanism. *The Journal of Nervous and Mental Disease, 74*
[52] Etheric body - soundessence.net. (n.d.). Retrieved April 18, 2021.
[53] Ibid

The etheric body is none other than an exact replica of the physical body. Each organ has its own etheric body. Their main function is to absorb light and heat from the sun – energy - and to transmit the energy to all parts of the physical body. The etheric body is the mold of the physical body, the archetype upon which the denser physical body is built.[54] In other words, and from a scientific point of view, this is a mere relationship between energy and matter and how they interact between them in the physical world.

When etheric body decreases, people experience loss of energy, fatigue and ailments. It is the same principle in science, if the mass increases in speed, it expands and when it loses energy it decreases. Alcohol drugs and other substances damage both the physical body and etheric body, and that impacts the mental and emotional health. Ailments in the physical body will reflect in the etheric body, that is why it's important to maintain good physical health, and good emotional health. Upon death, the physical body decomposes, and the etheric body leaves the body to be united with the energy field. Etheric body does not stand on its own without the presence of a

---

[54]The Astral and Causal Bodies. (n.d.).

physical body. [55] We are embodied with energy, and it is through our mind, and senses that we become part of a greater energy field, that of Universal Consciousness. Our etheric body as science describes, is an energy grid that enables us to function in a different realm than the physical world. The point is that we get to experience our etheric body through our meditation practices.

We are made of high frequency vibrations

God and quantum physics have become one of the most intriguing topics thus far in the area of spirituality. With humanity progressing, our thoughts have become evolutionary and people no longer follow dogmas without asking questions. As strange as this might sound, science and religion intercept. Since I am not a physicist, I will explain in simple terms what science has shown us concerning the physics of God. Quantum physics deals with light, waves, electromagnetic waves, atoms, electrons photons, in sum, all the components that are at the foundations of matter and energy. One of the questions that puzzles humans, and those who have an interest in quantum physics is identifying the energy that people refer to as *positive energy, the*

---

[55] 45 Metaphysical Foundations of Modern Physical Science, p. 275.1

source, good vibes or *bad vibes.* By definition, quantum physics is the science that studies the tiniest particles of matter, subatomic particles, the light the various waves and their frequency. Yet spirituality is as broad and as inclusive than physics, it ranges from astrophysics to quantum physics.

"Let there be light!" it says in Genesis 1:3 of the Torah. Light is the essence of life. There is light inside of each one of us. Light is energy and as such it is present everywhere. Quantum physics describe the light as a force of nature that combines electricity with magnetism. The smallest light particle is called a photon.[56] (Cited in Dennis Zetting)

David Blatner, in his book *Spectrums,* describes photons: "Photons – known as the carrier particles of electromagnetism – are literally what make our universe possible. Light – the oscillation of electromagnetic waves, the transmission of photons – is like the lifeblood of the cosmos, carrying packets of energy from one atom to another...."[57]

Thus, light comes in two states, as we call superposition, as particles and as waves. When light

---

[56] Zetting, Dennis; Zetting, Randi. A Quantum Case for God. Kindle Edition.

[57] Zetting, Dennis; Zetting, Randi. A Quantum Case for God. Kindle Edition. *Spectrums: Our Mind-Boggling Universe from Infinitesimal to Infinity,* Blatner, David. Bloomsbury, New York. 2014 (cited in Dennis Zetting pg 22)

engages with an electron, it manifests in the form of particles, when it does not engage with anything, it manifests as oscillating waves. Electromagnetic radiation or otherwise called as EMR is a spectrum of different light waves. This spectrum is comprised of different wave types: Radio waves, microwaves, infrared waves, visible light waves, ultraviolet waves, x-ray waves and gamma waves.[58] The light waves identified on the EMR exist throughout the universe. They are everywhere. Here on earth, we are engulfed with radio light, microwave light, infrared light and visible light. The earth's atmosphere protects us from exposure to gamma light, x-ray light and most ultraviolet light.[59]

The next two questions that intrigue people, are: How does quantum physics relate to the higher power – God? How do we experience God through quantum physics in a physical world? Well, so far, we have learned that the human body is made up of particles and energy which keeps the body bonded. Our universe is made of energy, as since energy and matter are interchangeable, then they are present everywhere. Etheric body as described by Newton is made by particles of energy that bond the physical body together. Even though Newton said that there

---

[58] ibid
[59] ibid

aren't so far any made known laws of physics that explain how energy behaves inside of the human body, we can yet acknowledge that it does exist. "Our personal two-dimensional energy body — variously called the astral body, light body, etheric body, spirit body, or subtle body — is the perfect, luminous counterpart of our physical body." [60] Our energy body is always with us and gives life to all our thoughts, feelings, and emotions. As I have mentioned earlier, these aspects of our body never die as we physically cease living. And this makes us immortal. [61] People project and absorb vibrations. This energy can be processed and magnified depending on the power of the thought – which releases waves in the form of frequencies. The stronger the thought, the higher the frequency. People manifest their energy in many different ways. Energy does not only manifest in thoughts, but also emotions of many kinds such as happiness, sadness, etc.

Joseph Selbie brings in his book the Physics of God a great example of how energy and mater works. Selbie states: "Energy can behave like matter, and matter can behave like energy. Pretty much, the

---

[60] Selbie, Joseph. The Physics of God (p. 173). Red Wheel Weiser. Kindle Edition.
[61] Ibid

physicists agree with this hypothesis that in a simpler term means: Everything we see, and touch has both energy and matter, and so is the human body, as well as the universe. Light can behave like waves or like particles. Atoms can behave as particles or like waves. Scientists call this duality a matter-wave. This is why matter and energy are inseparable, and this is also in congruence with Newton's observations. We cannot exist as a matter only without the energy Selbie continues.

1. Our physical body is interpenetrated at every level by the hidden-to-the-senses of the energy verse.
2. Like the universe our physical body is an ultra-high definition holographic projection.[62]

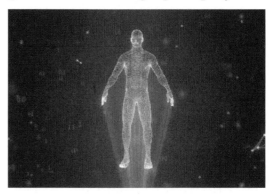

Image credit docwirenews.com

We, as humans – regardless of our physical existence in the form of matter, have our own holographic template or energy body. It has various names, such

---
[62] The Physics of God page 104

as astral body, etheric body, subtle body. The energy pattern in the astral body vibrates at a high frequency that is impossible for the human to see or touch. Whereas our physical body is made of lower frequency vibration, that are part of the matter. [63]

In Hinduism, energy is called Prana, in Chinese it's called Chi, in Greek as Pneuma, in ancient Egypt as Ka, in Christianity is holy ghost, or Holy spirit. Stillness, meditation rituals, yoga, breathing techniques, flow of energy, and also prayer are some of the energy practices. Just as I've mentioned in the beginning of the book, Universal Consciousness is a timeless, spaceless body of energy that we can reach through meditation. The entire physical universe exists as a relatively tiny, three-dimensional space-time-and-matter bubble immersed in an effectively infinite ocean of energy. The ocean of energy is two-dimensional and nonlocal: without time, space, and matter.[64]

The inner self seems to be able to perceive this vibratory energy image produced by neurons at the end process in the neurons of the prefrontal cortex. The inner self also seems to be able to produce mental vibratory waves of unknown frequencies by the mind

---

[63] Selbie, Joseph. The Physics of God (p. 105). Red Wheel Weiser. Kindle Edition.
[64] Selbie, Joseph. The Physics of God (p. 170). Red Wheel Weiser. Kindle Edition.

that causes physical corresponding resonant-like vibratory waves of energy in neurons of the brain. This way, the physical self and the inner self appear to mutually interact in the prefrontal cortex. This is the most common example that many people –perhaps - will identify with. [65]

The key question here that needs to be explored is how matter and energy came to be in a dual relationship with each other and how they bind together in eternity. This is perhaps the question of our existence. On a smaller scale, we always wonder about the harmony of matter and energy inside of us.

Life appears to be a manifestation of the constant subtle interaction of the wave-packets classically known as 'matter' with the underlying vacuum field. These assumptions change our most fundamental notions of life. The living world is not the harsh domain of classical Darwinism where each struggle against all, with every species, every organism and every gene competing for advantage against every other. [66]

---

[65] Chung, Sung Jang. "The Science of Self, Mind and Body." *Open Journal of Philosophy* 02, no. 03 (2012): 171-78.

[66] Karpouzos, Alexis. *Universal Consciousness: The Bridges between Science and Spirituality.* Athens, Greece: Think.Lab, 2020.

Thus, life just like energy is moving constantly, advancing and energy never gets destroyed. The Darwinian explanation of life and hypothesis of the "the survival of the fittest" has many flaws and has to do with the scarcity in the resources of survival and not the real nature of each and every one, the essence and the duality between energy and matter. It is more a behavior that is attributed to animals, rather than human beings and their intellectual embodiment: Alexis Karpouzos continues:

> *Organisms are not skin-enclosed selfish entities, and competition is never unfettered. Life evolves, as does the universe itself, in a 'sacred dance' with an underlying field. This makes living beings elements in a vast network of intimate relations that embraces the entire biosphere itself an interconnected element within the wider connections that reach into the cosmos.* [67]

Behavioral sciences explain human interactions and behaviors based on patterns. Spirituality explains certain behaviors based on the interaction between matter and energy.

---

[67] Ibid

Perceptions of energy in our thoughts.

People absorb vibes. Interestingly enough, sometimes people act like sponges: Certain people project good "vibes." They are always happy, and full or positivity. In a same way, some will project bad vibes, meaning, they are very toxic, or negative people. Many of us distance ourselves from toxic people. Yet we realize that negative people exist. From a social point of view and based on the set of shared values and norms, these people do not hold standards or morals. In the spiritual world we realize that they live in a different realm or dimension. We know that energy comes in [low vibration and frequencies] and [high vibrations and frequencies]. Even though in a different realm, these people carry low vibes that cause turmoil in others. The point of this illustration is to make the reader aware why certain people diffuse negative energy and where it stems from.

Imagine a wheel of dark energy. It can swipe and absorb everything in a black whole. The answer is easy and without reinventing the wheel. It all starts with a person's experiences in life. While this is not an excuse for people to be malevolent. We all recognize that some will be carriers of the dark energy. Life can be hard on some people. They grow up without love, without the care of their family. They can be product of many types of abuse, sexual, mental, emotional, physical, whether born into a family of wealth or poverty. It really does not matter. When positive energy is absent, and negative energy takes over, in the form of the abuse I'd mentioned above, the person does not know how to produce positive energy or convert the negative into positive. They continue to live their lives and make use of the dark energy they carry. Some even use that dark energy to advance in life. Advancing in life is part of human nature, therefore instead of magnifying positive energy, these people magnify the negative polarized energy. Our thoughts generate energy and vibrations. The size of positivity or negativity determines the frequency and the size of our thoughts. While you catch yourself living on the negative, it will magnify, and the opposite pole will do the same.

Etheric body and inner self

Inner self and etheric body are closely related to one another, and they are both considered as the soul, or Atman, yet there are differences, substantial distinctiveness between these two concepts. To put it in simpler terms – etheric body is literally a scientific concept and explained from a scientific point of view. A person is aware of the existence of the etheric body and later on when we discuss our relationship with the whole, we will see how our energy field connects with the universal or larger field of energy.

Inner self cannot be complete if there is no thought process. It is the mind that acts as a catalyst for us to experience the etheric body. The mind through our consciousness enables us to connect spiritually to the higher power and scientifically to enter the grid of the larger energy. These experiences are very unique and do not happen to everyone. In order for a person to enter earth's electromagnetic field, he or she must be skilled and devoted to meditation.

About seven years ago, I had three experiences where I was able to enter earth's magnetic field. When it first happened, and since it was a unique event, I was really scared, and I did not realize what was happening. As I entered the meditation, I sat in silence, and I experienced mindfulness. Then all of the sudden, I felt as if I'd entered a spinning swirl. I felt very lightheaded, I had a feeling as if my body was spinning with the speed of the earth. My etheric body had entered the larger energy field and the spinning increased, when in fact it was not me spinning. It was my etheric body that had come in contact with a strong field of bursting, swirling energy. I am not sure how long this experience lasted, but I am pretty sure it was not that long. I opened my eyes because I had to take control of my physical body- I closed my eyes again and the energy took over a second time. I stayed for a little bit, - again no sense of time – then I took control of my senses and continued with the meditation. We have to realize that in this unique experience, I did not get in touch with my inner-self or true self, since emotions were not involved. It was a burst of energy that absorbed the etheric grid into the electromagnetic field.

As I mentioned, in this particular case, I did not experience the inner self, this energy takeover was simply my etheric body acting "out of control" and wanted to join a larger energy field, that of the earth since the speed of me spinning was hardly controlled by me. In that moment, I felt as if I was never coming back to my physical self. I was only able to gain my control through my mind and since this was a new experience, I tried to understand for years what caused it and why was I part of this adventure. It wasn't until years later when I developed an understanding of inner self that I was able to realize that I had amazing experiences.

In the figure 2a I have tried to summarize in a very modest formula how I understand the inner self. As I've stated before, the inner self has two elements, and both of them are part of the subtle body. First being the etheric self, which is energy, and the other being the mind and the consciousness, a connector

77

to our physical body. As we saw before when we analyzed the etheric body, inner self when embodied personal characteristics, belongs only to a particular human being. For example, my inner-self even tough subtle, it is very personal, yet part of the whole – While the physical body dies, inner self dissolves from the body, and loses that identity, to later be reunited with the higher self. In simple terms it can be explained like this: When we sit in meditation, we are fully aware of our existence. We are in control of our mind and we can control our thoughts as well. The moment that we let go of the thoughts, the true self becomes self-evident and we experience just energy – the etheric body. People might experience the subtle body differently. While the subtle body interacts with a larger field of energy surge, one can experience a feeling of being surrounded and in the eye of a whirlpool or rip current. This feeling intensifies if a person meditating holds crystals – such as amethyst in both hands. This practice is very beneficial to also cleanse your chakras. Thus, in this case we are bringing our etheric body to a point that we are enabling interacting with energy. This practice is also great for healing, yet people need to be skilled to perform such a task. In a much simpler way, it will revitalize the energy grid and raise your vibrations.

Inner-self = etheric body + consciousness.
$$(is) = eb + c$$

Figure 2a

Thus, in this chapter, I've discussed a few concepts to open a window of information. As I try to break down the ontology and the relation of energy and matter, I explained that we are made of both, and we cannot exist if energy or matter is missing from our "being." The cosmic breath we intake – prana, or pneuma gives life to every single living cell in our body, and our physical body is the perfect organism to enable such magnificent marriage between energy and matter.

THOUGHT AND MIND

*"O God Almighty Lead our mind toward the virtuous path" Vedas (*Rig 10.35.9)

As we navigate through this book, we come to realize that knowing the inner self would be absurd without the power of the mind. From a cognitive standpoint, it would be pretty impossible to complete a thought and think critically without the presence of "the physical organ" in charge of the thinking process: the brain. The book does not focus on the brain's anatomy but touches upon some critical parts and functions that enable the thinking process and, therefore, the delightful meditation process. We need to understand that we need a connector to tie the physical body and inner self, and as I mentioned above: It is the mind that makes that connection.

For example, many enlightened people believe that the mind is a vital – yet subtle part of our inner self. That is undeniable. Let's take the case of a person with an impaired brain. If the brain's primary functions are impaired, the person cannot think at the age-appropriate level, have reasoning skills, or higher thinking skills. Many people whose brains have suffered physical damage - whether innate or caused by trauma, cannot even take care of themselves and be independent. I am not saying that people with brain and mind impairment do not believe in God. Many of them pray to God and significantly accept God's existence.

In a literal sense of the word, the mind is crucial in one's life and survival. Although while the mind in our daily lives is responsible for completing tasks from basics to more complicated, the mind is also the tool that makes the connection between the inner self and God. In the diagram below I have explained the process of integrating the mind to make that divine connection between the physical body and the Universal Consciousness or even Atman.

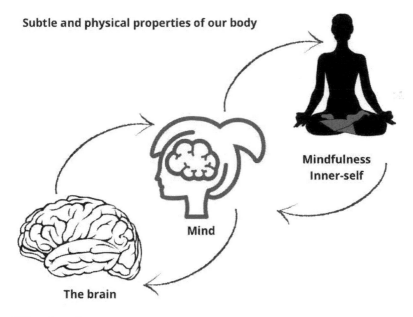

The brain

The brain is the body organ responsible for all of the vital functions. The brain regulates the biological and cognitive systems in our body, starting with the nervous system, respiratory, circulatory, sensory, and of course cognition, and many more functions. When the brain dies, the body dies and vice versa.

One of the systems widely discussed nowadays in spirituality is also the sensory system, apart from cognition. The senses are like the messengers that connect the brain in our body to the world by sending information through hearing, smell, touch, sight, and taste.

Thus, the brain receives information through our five senses: sight, smell, touch, taste, and hearing - often many at one time. It assembles the messages in a way that has meaning and can store that information in our memory.[68] The brain also controls our thought process, memory, and speech. Most importantly, it controls the movement of the arms and legs and other functions of many organs within our body – through the nervous system. That is how the brain sends signals and receives signals. The brain has three main parts that have a very distinct function.

---

[68] Mayfield brain & Spine. (n.d.). Retrieved April 19, 2021, from https://mayfieldclinic.com/pe-anatbrain.htm

1. Cerebrum
2. Cerebellum
3. Brain stem

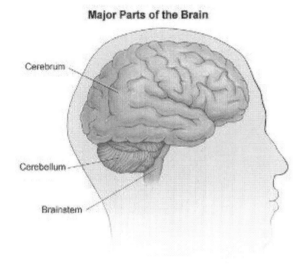

Photo Credit Hopkins medicine.org[69]

The cerebrum is the largest outer part of the brain. The cerebrum controls the thinking process, reading, and learning. It also controls speech, emotions, and planned muscle movements like walking. The cerebrum is also responsible for sensory functions. It controls vision, hearing, and other senses.

The cerebrum is divided into two cerebral hemispheres (halves): The left hemisphere controls the right side of the body, and the right hemisphere

---

[69] https://www.hopkinsmedicine.org/pedersen-brain-science-institute

controls the left side. The hemispheres are further divided in four sections, otherwise known as lobes. These are the frontal lobe, parietal lobe, temporal lobe, and occipital lobe. Each lobe controls specific functions. For example, the frontal lobe controls personality, decision-making and reasoning, while the temporal lobe controls memory, speech, and sense of smell.[70]

The cerebellum

In the back of the brain, the cerebellum controls balance, coordination, and fine muscle control (e.g., walking). It also functions to maintain posture and equilibrium.

The brainstem

The brainstem, at the bottom of the brain, connects the cerebrum with the spinal cord. It includes the midbrain, the pons, and the medulla. It controls fundamental body functions such as breathing, eye movements, blood pressure, heartbeat, and swallowing.[71]

---

[70] Mayfield brain & Spine. (n.d.). Retrieved April 19, 2021, from https://mayfieldclinic.com/pe-anatbrain.htm

[71] Kieffer, S. (2018, December 03). How the brain Works: Johns Hopkins Comprehensive brain Tumor Center. Retrieved April 19, 2021

Theory of mind

Scientists argue that the theory of mind is related to the frontal lobe. According to research in psychology:

"Theory of mind, the ability to make inferences about others' mental states, seems to be a modular cognitive capacity that underlies humans' ability to engage in complex social interaction. It develops in several distinct stages, which can be measured with social reasoning tests of increasing difficulty."[72]

The same group of researchers argue that theory of mind first appears when the child is approximately, eighteen-months-old and, can do tasks such as joint attention and proto-declarative pointing, and so forth. Thus, there are a few stages of mental development from toddler to teen. For example, by the age of two a child is able to show a desire – wanting a toy or a hamburger. By the age of three or four the child is able to understand that other people can hold false beliefs. By the age of six and seven the child becomes aware of other's mental state, or second order false beliefs. Later, between ages nine and eleven children develop the ability to understand and recognize faux pas. A faux pas occurs when

---

[72] Stone, V. E., Baron-Cohen, S., & Knight, R. T. (1998). Frontal lobe contributions to theory of mind. *Journal of Cognitive Neuroscience, 10*(5), 640-656. doi:10.1162/089892998562942

someone says something that they shouldn't. Not knowing or not realizing that they should not say it.[73]

Studies also show that children with autism, Down syndrome or other mental impairments, do not fully develop the mind. In the case of children with mental impairment, their development is fragmented or impaired in one of the stages shown above. "Rather, their errors may be due to problems connecting their theory of mind inferences with an understanding of emotion and other metacognitive skills-beyond the first and second order." [74]

Theory of mind does not only allow space for ourselves to create our own beliefs and desires, but it enables our mind to understand that others also have unique beliefs and desires that are different from our own. By realizing this competency in ourselves and others, we are able to engage in daily social interaction as we interpret the mental states and infer the behaviors of those around us (Premack & Woodruff, 1978).[75]

---

[73] Ibid

[74] Stone, V. E., Baron-Cohen, S., & Knight, R. T. (1998). Frontal lobe contributions to theory of mind. *Journal of Cognitive Neuroscience, 10*(5), 640-656.

[75] Ruhl, C. (2020, August 07). Theory of mind. Retrieved April 20, 2021.

Philosophers and psychologists are concerned with the duality mind and matter, where mind is a process where the brain is the matter responsible for this process. Apart from the physical and material needs, the mind is a medium that serves as a liaison between the inner self and God or the Universal Consciousness.

That is where the info is processed and through the mind, we are able to gain knowledge.

The mind, the senses, and the enlightenment

Knowing that the mind is what connects us to the Universal Consciousness and knowing that the mind has a functioning role in our daily life, we realize that the mind is necessary for both processes.

In Vedanta – it says, by the mind alone the enlightenment can be grasped or realized says Swami Sarvapryananda.[76] And yet he continues, all the knowledge that we gather from our senses is processed in the mind. *Pure consciousness appears in the mind. The mind is like a mirror. The mirror will reflect the sky and the mind reflects the consciousness, and the consciousness illumines the mind. Little particles of electricity circulate in the mind*

---

[76] Vedantany1894 (Director). (2021, January 14). *Ask Swami with Swami Sarvapriyananda | January 3RD, 2021*

–meaning, the mind absorbs and produces energy. We also learned that through quantum physics. This transportation of energy from outside to the mind is done through our senses. Thus, objects are connected to the senses, and senses are connected to the mind. In this case the senses serve as portals, conduits of energy.

These processes allow us to activate our knowledge and create images of something that we do not see, for the moment, yet we have seen in the past. For example, if we have seen a colorful singing bird in the past, the mind has stored that image and the movement our mind asks for mental image of that bird, the mind creates it. It is through the senses that the mind creates that reflects that very same physical object, bird, flower or table in our mind, and it can even create new images just based on the particular image. [77]

The activity of the mind is necessary for any type of knowledge. To generate the knowledge, we need (reflected) consciousness. The ultimate goal of spirituality is to experience and realize the unlimited consciousness (Brahaman) or God. To realize Brahman, people need to study about it, meditate

---

[77] Swami Sarvapryananda uses a metaphor of a flower and how the mind reflects that image or how the mind creates reflections of other images for that matter. The idea is that the mind is like a mirror. It reflects the consciousness.

and pray. However, the mind has a very important role in enlightening us, and that according to Swami Sarvapryananda can be explained with a metaphor: The sun, moon and earth:

Imagine is the sun is Atman, pure consciousness, the moon is the mind and the earth, is simply the world. The sun illuminates the moon, and the moon illuminates the world. We all pretty much agree that to enlighten ourselves, we have to receive light from the reflective consciousness. In the case of the eclipse, the sun shines on the moon, yet the moon does not shine on the earth, instead it reflects the light back to the sun. In this moment, -Swami Sarvapryananda explains- the enlightenment process and pure awareness happens. For the mind to be fully enlightened, it must be detached from the physical world. We can achieve this state by meditating, through prayer, signing mantras and many more spiritual practices.

The conscious mind

Our existence on this planet or universe is a question for the ages – it has puzzled humans since they became aware of their existence. Granted, the only way we know we exist, and we are alive is because we have a conscious mind: We think. Our conscious mind stems in our brain as such and is a very subtle yet individual concept. The way we think, and operate, makes us different from one another. It separates our thoughts from the rest. The irony is that we are not able to read each other's minds either. Explaining how the conscious mind works is not as easy as it may sound. It remains a mystery for scientists and philosophers as they have different postures on the subject. On the other hand, mind and thought are part of cognitive and metacognitive disciplines. Various underlying and direct assumptions try to explain the conscious mind by analyzing the human behavior. Following the same complexity of explaining the conscious mind, one can deduce that it is a physical process that enables a cognitive function and enables memories and other mental processes to bind together coherently. Hence, the conscious mind stems from the human brain and is a product of the human brain. Animals do not have a conscious mind. Our brain is one of the most vital organs of our body that is responsible for many physiological processes, organ function, yet explaining how the brain generates the conscious mind is yet a mystery. The brain consumes 25% of our energy, and it is only 25% of our body weight. Researchers and modern medicine have tried to study what our brain can do and how it changes upon receiving information. These studies include the use of MRI and imaging. One of the questions that puzzles the scientists is how we create thoughts and memories, and therefore explain our existence. As we have seen in the previous sections, the brain is divided into lobes and parts with specific functions. The part that controls the thought process is the cerebrum. More importantly, from the experiments conducted, if that part of the brain does not function properly, then the conscious mind will not perform. Brain researchers and psychologists have come up with many theories. Still, one of the most influential is the Binding theory – which tries to explain how the brain bonds together all the processes and the information. Cognitive psychologists often explain the answer to the question: How does the brain integrate everything into the conscious mind? This process is called "the binding agreement," meaning bringing all the brain regions together. What is more important to understand about the conscious mind is that we are equipped with a mechanism to understand the world around us and expand our learning and understanding beyond what we see, touch, and how we reason. It means that we can presume and hypothesize various theories that can explain our existence and our universe since its conception. Thus, if our brain is impaired, then the cognitive process is fragmented. Therefore, the conscious mind is also fragmented. Our brain collects information from our sensory organs, and based on that information, our brain creates and updates models and schemata to analyze new data.

A healthy mind mirrors a person's behavior, lifestyle, and decisions. Cognitively, the mind is able to balance the processes, whether caused by intrinsic chemical reactions, or by the ability to receive signals and stimuli from the outer world. Our mind does so, and it processes them into thoughts.

Lastly, the mind is responsible for generating thoughts. It is the computer where thoughts are created and are processed. We make use of our minds as the only tool that enables us to not only be aware of ourselves and who we are but also to function as regular physical beings. As we saw in the chapter, when the brain, is impaired, thoughts are impaired and physical and mental abilities are limited. Our consciousness operates differently than the inner self. While our consciousness is product of our brain, inner self is our subtle body. Yet the mind is the tool that can connect us to both, our consciousness and our inner self, and therefore the Universal Mind. While we enter a meditation session, our mind directs us to our consciousness if we are not skilled enough to go there.

The conscious and the subconscious mind

The conscious mind is our voice of reason. It directly impacts societal values, the external environment, and at the same time, affects us in our decision-making. On the other hand, the subconscious mind is quite mysterious. It harbors all habits, emotions, tendencies, memories, experiences, and self-image – that people often don't realize. Subconscious is where people store everything that impacted them as humans in their present life. These memories are direct experiences once lived and now stored in the dark corners of our subconscious mind. Even if they seem to fade, they still exist. Such are bad memories from childhood, trauma, or even a happy moment. There are two levels in our mind — the conscious or rational level and the subconscious or irrational level. People think with their conscious mind, and whatever they believe habitually sinks down into the subconscious mind, which creates according to the nature of thoughts. The subconscious mind is the seat of emotions and is the creative mind. This is the way your mind works.[78] The spiritual writer Joseph Murphy says:

---

[78] Murphy, Joseph. The Power of Your Subconscious Mind (p. 12). Pandora's Box. Kindle Edition.

"Your subconscious mind accepts what is impressed upon it or what you consciously believe. It does not reason things out like your conscious mind, and it does not argue with you controversially. Your subconscious mind is like the soil, which accepts any kind of seed, good or bad."[79] Child trauma and bullying can impact a person if exposed to violence and physical abuse at an early age. Also, criticism and narcissism will leave a scar on children's subconscious minds. Such impact alters their self-esteem and decision-making power once they grow older. These experiences will later affect the inner-self and that person's reality and spirituality. Thus, the inner self is very close to the subconscious mind, and as a result, they are both parts of the subtle body. Joseph Murphy continues: If you consciously assume something as true, even though it may be false, your subconscious mind will accept it as true and proceed to bring about results, which must necessarily follow because you consciously assumed it to be true.[80]

---

[79] ibid.

[80] Murphy, Joseph. The Power of Your Subconscious Mind (pp. 13-14). Pandora's Box. Kindle Edition.

Emotional self

When we talk about emotions, we are aware that they trigger reactions inside of us. Reactions can have many responses. The emotional self is the way our mind reacts to external stimuli. Psychologists and people in the field distinguish two types of emotions: Positive emotions and negative emotions. Generally speaking, emotions reflect the way we feel, how we react, and how we handle our feelings. In the previous section, we reiterated that emotions are part of our subconscious mind. It is important to understand that the external environment triggers our emotions, yet it is us, our will, our reaction that accounts for our responses. Our emotional self is also part of our subtle body. It is quite ironic that we have strong feelings and emotional reactions, yet we cannot see our emotions; we can only see their reflections in our attitudes and behaviors. In the same manner, our emotional self is subtle, yet it can be fragile, delicate and very tangible. Even though we cannot touch or see our emotional self, we realize how we feel and express ourselves a certain way. The physical organ responsible for processing our emotions, without a doubt, is our brain. During the thought process, many emotions are manipulated in the brain, whether knowingly or unknowingly, but incited by stimuli. Emotions are not possible without the presence of the mind, yet they have a mind of their own. They reflect our mental state and behavioral health. Emotions will also trigger various types of behaviors and determine in what mental state a person is while he or she experiences an emotionally charged self. Our emotional self is closely connected to our self-consciousness and impacts decisions based on our preconceptions and values. If emotions are triggered, people will most likely react to shield themselves from getting hurt or protect their inner self from getting hurt.

Emotional pain is one of the mental states that people have a hard time coping with. It is pervasive, it makes people react poorly. It brings sadness and sorrow. Many have a hard time bouncing back from a painful emotional event. Others make irrational decisions based on emotional distress. Emotions are triggered by many including politicians to advance agendas, and by corporate to sell, ideas and products. Marketing strategists know how to tap into that emotional self and have shoppers or consumers buy emotionally and even eat emotionally. Emotional impairment, as I've mentioned earlier, impacts our inner self in a way that once we can be scarred, it causes a disconnect with our spirituality, and impedes opening the door to Universal Consciousness. Understanding our emotional self is paramount and people tend to be resilient in the majority of cases because we all have the ability to use our own defense mechanism and survival mode. Deterrence of situations where we know we'll be detrimental to our emotional health is important. Simply said, it is wise to walk away sometimes than to confront.

The emotional self is widely discussed in self-healing techniques. Life coaches reiterate the importance of healing. Before anything, soul healing must take place for a person to start restoring his or her life. The emotional self, on the other hand, is quite different from the inner self and self-consciousness. When emotions take over the thinking process, they will impair and distort the reality to satisfy the emotional needs a person has or alleviate the emotional distress. People who have been emotionally abused also have a negative perception of themselves and the reality in which they live. This perception impairs the ability to understand our inner self. Many people who have been emotionally abused in their childhood have perpetual hate for themselves and their image that needs to be addressed. Sadly, many individuals who do not seek help or attempt to help themselves become emotional abusers for their children. Data and experiments in psychology reveal successes in the field of psychotherapy, and lately with meditation techniques.

Conscience

One of the concepts that is key in understanding the dogma behind religion is the conscience. In the previous chapters, I've mentioned that we have a social identity which makes us part of a group, and since the group, community - starting with the nucleus "the family" - sets values, it makes the conscience one of the moving forces of our society. Religion has used this highly cognitive process to propagate its dogma, in order to keep the followers in line. Of course, there is a good side of it. Yet while it crosses the boundaries of individual freedom dogma becomes a controlling mechanism, then it becomes highly abusive, drifts away from spirituality and God, and turns into a cult.

People become aware of their conscience at an early age, which later in life serves as a precursor to develop a consciousness. Hence, our human mind is well refined and malleable. We develop many learned behaviors, and thanks to the power of the mind we create systems of values and beliefs that later in life become a pivotal force in decision making, and a base for religious or spiritual practices. Many behavioral theorists argue that children become aware of their conscience at the age of four or five. Their parents or caregivers teach the children at this age moral values. By the age of nine, they start developing their own conscience, and around twelve, they become more refined and sophisticated – they hold a set of values. Thus, in early childhood and later on during the grammar school age, children face decisions; conscious decisions, in other words, how to tell the right from the wrong and many moral decisions. Sometimes children can be brutally honest. Many behaviorist points of view argue that these preconceptions are taught at an early age. In my personal opinion, and as I have stated elsewhere in this book, these trained behaviors have to be in alignment with cultural upbringing and do not necessarily reveal the inner self that we often confuse with our conscience. Hence, the inner self and the conscience are two different things that I would like to discuss and explain to my understanding. To understand the inner self, we must detach it from the physical body, and the conscience.

Conscience teaches people how to act and deal with others in the physical world. The very nature of conscience is subtle. Conscience is an intrinsic process of our mind, yet it is also heavily influenced by the extrinsic environment. One might ask about the difference between the conscience and the inner self. There are many differences between the two; the most important thing is that conscience makes us human, very distinct from animals. Inner self makes us divine therefore it connects us with the Universal Consciousness. Here is the schism between conscience and inner self. Inner self is part of Universal Consciousness, and self-conscience is very individual. It is unique and distinct but is a product of a certain upbringing, culture, tradition, and environment. At the same time, inner self is our connection with the divine, God, and the source. The inner self is pure and flawless. It is "the better me." The self-conscience construct has many flaws because outside factors prompt it. Since the individual is part of that particular group identity – whereas social, religious or class identity, the self-conscience must align with that specific group's norms and rules.

The thought and its power

Chatterjee (2004) argues that we have two selves. One is the conscious self which actualizes the thought and action, and there is another self which works a guide and forewarn us. Many people often confuse the little voice inside of them- with the inner self. Yes, conscious self and inner self, are very close and often overlap, yet they are very distinct. While we experience inner self in a meditative state, we can experience our consciousness in a normal state. Usually people who meditate have the ability to let the consciousness go and surrender to divine forces or energies to seek for answers or ask for wisdom. Whereas Chung distinguishes between inner self and consciousness. The catalyst in this process the mind plays as a thought process yield by the brain. Chung argues that the conscious mind is not the inner self. I agree with her argument. Realizing that the conscious mind is a product of the brain, in conjunction with the output from the environment.

The power of thought is more than we can imagine. The thought never dies – by that I mean literally and spiritually. Many philosophers left their legacy which is in fact the product of their thinking, and they will be referred and remembered for their thinking. We think and as we think we produce an abundance of thoughts. It is estimated that a person's mind processes between 50,000-60,000 thoughts per day. [81] That is a tremendous number. Most of the thinking process is done naturally and we are not aware of our thinking unless we direct ourselves to

---

[81] Sasson, R. (2021, April 17). Concentration exercises for training and focusing the mind.

a particular thought. But the rest is done unknowingly. With this magnitude of processes and thoughts our minds move rapidly, and the pressure builds up if the mind is not given a break and started anew.

Quieting your mind: Find your inner peace

Finding inner peace is very meaningful for someone's spiritual growth and in the way of experiencing Atman, or inner self in harmony with God almighty. It is the mind what keeps us fully aware of our existence, the relation with the divine and the Universal Consciousness. Finding inner peace comes with quieting the mind. [82]

Peace of mind is the status of mental and emotional serenity. It entails stillness, and relaxation. When the mind is quiet and people experience a feeling of exhilaration, they experience a sense of happiness and freedom within.

It is hard to achieve inner peace, or peace of mind. With 50,000 (average thoughts per day), it requires focus, concentration and detachment from life's worries. That explains why meditation is not an easy process. Once the mind is quiet, awareness takes over and it paves the way for inner peace, bliss, and unification with the higher power.

---

[82]

I will bring my personal experiences to illustrate how to attain peace of mind.

About eleven years ago, I started graduate school. I was enrolled in a Doctorate program. I was very happy, and it was a dream come true. I made it from a large list of candidates, and I was picked because I had scored well on the math portion of the GRE. It was during a time when they did not beg or recruit people to go to grad school. It was based on personal and academic achievements. When we started the program, the stakes were high, the reading became overwhelming. The reading material for the week totaled over one thousand pages including theory, books, articles etc.

My brain felt like it was exploding. I started dreaming about my classes and I became very worried. If my mind was working intensively during my sleep, I realized that I needed to take another approach. It was then when I had discovered meditation, and its different techniques. I always thought I was closer to the divine.

Every now and then I had to cleanse my mind from all that thinking process, all the case studies, literature review. Fortunately, I live close to the beach and I get my strength from the majestic view of the ocean, and the feeling of being close to the creator. While I sit on the grass, sand or the rocks and gaze at the water, my body is engulfed with a sense of happiness. It is the sort of happiness that the heart rejoices. It is not happiness from life accomplishments, instead, it is a feeling that I was chosen to have a physical body and physical life, to be part of this great journey. As I sit at the beach and mediate, it is then when I feel that inside, I am my true self, an inner self that is completely separated from my physical body. A true self that is experiencing bliss.

In this state, my mind is quiet, it is occupied with positive energy, thoughts are gone.

I sit quietly and close my eyes. I erase everything from my mind. In that moment nothing stands in the way, between me and the Creator. I experience the calmness that nature allows me to experience, I breathe freely and as I close my eyes, all the heaviness in my mind disappears, and with it, everything else that entered my mind from our daily lives. Everything is forgiven, dissolved, and turned into sparkles back to Universal Consciousness.

Tip:

> *You can visualize yourself doing the same thing. If you cannot go to the beach, bring the beach to you through visualization.*
>
> *Sit quietly in the comfort of your house, or out in nature. Look around you and find something*

that will bring happiness. If you are in nature, it should not be hard. Nature is the best home the Creator has granted for us to enjoy and experience. Gaze far away into the horizon, or contemplate the trees, anything that is pleasing to the eye. Once you engulf yourself with the feeling of love and gratefulness, then you have emptied your mind from all that array of information, influence, confluences with other people, media, social media, co-workers, friends and family. In that moment it's just you, the nature and the Creator. Continue to empty out your mind of unnecessary things. Think of it as a recycling bin that needs to be emptied out one by one. Fill up your mind with radiant light, energy, and thankfulness for getting your peace back.

THE UNIVERSAL CONSCIOUSNESS

Otherwise known as the Universal Mind. Many people are confused about the Universal Consciousness and its relation to the physical world and precisely how it affects us. In daily practices, the Universal Mind, is otherwise a data base, where all the information related to the physical and spiritual world is stored. One of the most effective ways in explaining and understanding the Universal Mind comes from Dr. Gautam Chatterjee which states:

The Universal Mind is not separated from us and we are not separated from it. We have to shift our awareness self (ego) before a true stream of real data can go back and forth between our conscious mind and the Universal Mind.[83]

Studies reveal that energy is everywhere and our thoughts move energy. In his book *The Physics of God,* Joseph Selbie, explains that people relie on modern science and technology to explain various

---

[83] Chatterjee, G. (2004). The Universal Mind timeless and Ageless. *You Can Change the World.*

phenomena. Humanity relied on spiritual individuals such as shamans, priests, mediums and other people who were able to connect with the Universal Consciousness and perform healing rituals. Chatterjee sustains that, "Regardless of physical distance, early civilizations had many common characteristics that were a result of direct connection with the Universal Mind." In addition, Chatterjee suggests that to access the Universal Mind, one must shift his or her awareness from (self) or ego and enter a trans state. What is key in this process is the use of our mind as the medium to go back and forth between the conscious mind and the Universal Mind. In the previous chapters, I analyzed the role that the mind plays in relation with the physical and spiritual world. However, something that I can reassure with certainty: The mind can be deceiving, manipulative, driven by the ego, or egocentric tendencies, yet when the awareness is shifted to the spiritual world, the mind loses its powers, and is stripped from all impurities. The mind cannot "cheat" the Universal Consciousness, or the eye of the world. It is that simple.

The Universal Mind is pure knowledge. In contrast with the individual mind, the Universal Mind – the place where we try to organize, classify, layer information cannot be manipulated. Its omnipotence relies on the ability to be true, pure and available to those who know how to get in touch with it and benefit from it.

The "Universal Mind" (or the global consciousness as it is differently called nowadays), namely the totality of all the actions and thoughts, is what creates reality around us in the same way quantum physics teaches us that the observer chooses his reality among an ocean of probabilities the moment of the observation, i.e. at the act of thought.[84]

People usually are driven into visualizing everything into their physical understating of the world, and while visualization is part of a higher consciousness processes, the Universal Mind, is not physical and nor can it be. The Universal Mind is ageless, timeless, spaceless. It does not occupy a certain space, it is the space, it does not occupy a certain time, it is the time, it does not have a beginning and the end, it is the infinity, it does not have physical boundaries. In Buddhism, the Zen teaching of Bodhidharma states:

This mind, (Universal Mind) through endless kalpas without beginning, has never varied. It

---

[84] Karpouzos, Alexis. *Universal Consciousness: The Bridges between Science and Spirituality*. Athens, Greece: Think.Lab, 2020.

has never lived or died, appeared or disappeared, increased or decreased. It's not pure or impure, good or evil, past or future. It's not true or false. It's not male or female. It doesn't appear as a monk or a layman, an elder or a novice, a sage or a fool, a buddha or a mortal. It strives for no realization and suffers no karma. It has no strength or form. It's like space. You can't possess it and you can't lose it.

Its movements can't be blocked by mountains, rivers, or rock walls... No karma can restrain this real body. But this mind is subtle and hard to see. It's not the same as the sensual mind. Everyone wants to see this mind, and those who move their hands and feet by its light are as many as the grains of sand along the Ganges, but when you ask them, they can't explain it. It's theirs to use. Why don't they see it?

... Only the wise know this mind, this mind called dharma-nature, this mind called liberation. Neither life nor death can restrain this mind. Nothing can. It's also called the Unstoppable Tathagata, the Incomprehensible,

the Sacred Self, the Immortal, and the Great Sage. Its names vary but not its essence.[85]

Inner self and the Universal Mind

For us to get in touch with the Universal Mind, we must go through our inner self, and to do so we have to use the power of our mind. Meditation is beneficial to let go of all thoughts and preconceptions. Thus, by avoiding the conscience, the morality that characterizes, we can get in touch with the Universal Mind and the divine. In the meditation chapter I will share some meditation practices on how to escape consciousness and let go of fears, judgments, insecurities and inhibitions.

This book is not about the law of attraction, or other speculative theories about wealth, and glory. It is a simple way to start the spiritual journey. It is up to the reader to use this knowledge to his or her benefit. My aim is to also bring awareness to the struggles I had personally went through. Such as being an enthusiastic person all my life and how toxicity can overwhelm you physically and emotionally if not dealt with. I feel strongly about women, the elderly

---

[85] Chatterjee, G. (2004). The Universal Mind timeless and Ageless. *You Can Change the World.*

and children who go through emotional pain. These are the most vulnerable people in our society.

For example, once you crush a child soul through verbal, physical, or sexual abuse, that child is scarred for life. That child will hate their *inner self*. That child will do anything they can when he or she grows up, to forget the *inner self* and have a major identity crisis, get rid of depression, anxiety and even suicidal thoughts. Thus, the inner side of me is God in us, is the innocence in each and every one, and once it's taken it will break someone's soul. But the example I made about children who are abused, are not necessarily the only example where a crushed soul causes an identity and existential crisis. A person wounded deep in their soul has a hard time connecting to the Universal Mind. However, with help, support and warmth from loved ones – friends, family a person is able to restore faith and get onto the spiritual path.

In life we will always have blockers of energy, those who are soul crushers. We have to realize that negative people embody negative traits including energy. Once they absorb you in their vector, they will never voluntarily let you go. While you surrender, in that trajectory, your soul is crushed, your inner self is destroyed, that confidence dies. You become a soulless person, a person that just exists, but dead inside. Imagine that! Who can live like that?

Luckily, we are equipped with a defense mechanism to fight back. But not everyone is as strong. Many people do not make it out of that vicious cycle until is too late. And in that case, they age before it's time and they can never live a quality life, because, peace inside of them will never restore – unless they choose to do so.

We cannot deny the fact that the inner side is the promoter of the decisions that we make in life. It is something that we cannot avoid dealing with sooner or later. We can choose to remain silent, avoid all contacts, as a matter of fact when external factors are the biggest contributors in someone's life, it deals directly with how the inner self is impacted. No outside contact is ever left unnoticed. It might not impact us with the same magnitude, but it absolutely leaves a trace in our minds. Even a quick chat at the supermarket with a stranger, will trigger some reaction in our brain.

The inner self is as important as our vital organs. Without a strong inner side or inner self our lives can never be fulfilled and function normally. Although function normally does not mean to have an excessive social life, it means in fact to function

The outer side of me or the superficial me

The superficial me does not mean the shallow side of me as much as it means what I chose to show to the world; what I am, or what I am not. Showing the true self, sometimes means vulnerability. In some cultures, people find their true self later in life after their children are grown and there are no societal constraints to follow. For example, in Hinduism, this stage is achieved in the Moksha stage of life. Some devotees take their last spiritual journey and give up all material life to pursue a spiritual life. Imagine Mahatma Gandhi when he vowed to dedicate his life to his God and moved to the Ashram where he'd spent time making khadi.[86] When we consider his decision, in our Western culture, we cannot fathom the thought that we can give up everything we own and live with a few material things, yet an enriched spiritual life. The paradox here is striking. More than ever, our society is guided by superficiality. We pick and choose what to show to the world, whether in person, or virtually. The superficial side can sometimes reflect who we are inside, and this is achieved when we find our true self, but in the majority of the cases our superficial self is merely a distortion of who we are and how we feel inside. This

---

[86] Indian textile.

is an expression of unhappiness and emotional distress.

People have a hard time showing their true self, because of stigma, they do not want society to perceive them in their natural state, yet they put on a mask and become a persona that has nothing to do with their individuality. With the pressure from the society, people tend to change who they are inside and become pleasers of the crowd. Pleasing the crowd gives people instant pleasure and gratification, but in the long term, it deepens the identity crisis and it contributes to depression, and social isolation.

Our immediate environment

Home is known to be as the ultimate shelter for a human being to find peace and tranquility to unwind, but most of all to feel safe. Home is a necessity, but it can be privilege to some. Let us focus on the majority of people who have a shelter they call home. Generally speaking, we tend to decorate our home to our own taste and people spend tremendous amounts of money trying to make a house or an apartment feel like home. Home should be a place that makes us happy. If we cannot find happiness at home then, it is in the human nature to evade by escaping that environment because it does not fulfill the purpose therefore, it does not feel safe. In that case home fails to fulfill the emotional needs. How many people cannot find happiness at home? Whether unhappy spouses, teenagers, or even roommates seek to find an alternative place if their immediate need for peace, tranquility, emotional need and self-expression is not met. One of the reasons unhappy spouses leave is the hostile environment at home. An angry spouse no longer has an interest in the marriage or becomes argumentative because he or she is deeply unhappy. That is a warning that the marriage is becoming rocky and it needs to be dealt with in a proper way, either through counseling or through going separate ways. Children do not feel safe at home when they are not safe physically and emotionally. In that case children become victims of child abuse and neglect. Teenage children are a different story. Somewhere between a child and an adult "state of mind" teenagers will runway from home if the environment becomes unwelcome and the lines of communications with parents or guardians are broken. Thus, the first place we seek happiness is home. We need to be individuals more so than fit in the larger group so we can find this individualism ah a place we call our own: home. The need to be fulfilled individually starts at home.

The physical world and the greater picture

Imagine yourself sitting on a bench near the water. Next to you, there is an old tree that had just blossomed. As you sit, you stare at the horizon, and the sky is crystal clear. The sun reflects on the water, and the bouncy waves shine lively. You are contemplating the view, and you immediately think of the Creator. Nearby, a few seagulls fly and settle on the wet sand in front of you. They are aware of your presence and keep you company. The colors are contrasting, green, sandy, dark blue, and light blue. A sea breeze caresses your face and plays with your hair. You take a deep breath. You are happy to be alive. Again, you thank the Creator for making you part of this earthly experience.

This experience is not fiction, it's part of my life especially in the summertime. I go to the beach to meditate and while I settle down, I become aware of the environment in which I have been placed and I am part of it. The physical world is the surrounding world that we "as individuals" do not have control over, but nature does. Of course, there are esthetically manmade changes that we as citizens and constituents like to create as a relaxing environment, but those are touches here and there and they cannot override the natural beauty of our world. We have to establish a loving relationship with nature. Everything in nature carries energy, mostly positive energy. Here I am not concerned with forces of nature and their destructive nature, but the physical environment that we need to strengthen or body, and our mind. Harmony with our surroundings is a great way to find peace. Often, we use the outdoors to walk, to work out, yet many do not take the time to sit down and get in touch with it. Feel the energy, warm your body and soul with the sun's rays. Nature is the greatest diffuser of positive energy coming directly from the sun, and for that matter, the healer. At times, when I go to the beach, I see people in a hurry to finish their two-mile walk, yet I hardly see anyone take time to sit down and observe, contemplate, pray or meditate. It seems that people are always in a hurry to get somewhere. There is no benefit in the walking except what the doctor tells you - to pump blood into the heart and the organs. The importance of the environment in our spiritual awakening is paramount. When opportunity is given, people must benefit from it. Finding harmony and being in alignment with the source is beneficial to one's mental and physical health. In the upcoming chapters I will describe some of the techniques for spiritual awakening and meditation practices.

THE PHILOSOPHY BEHIND SPIRITUAL PRACTICE

When your heart rejoices, you experience a subtle feeling of worthiness. That is the good karma engulfing your heart with happiness. People usually have the perception that karma is pernicious and that it will get to you sooner or later. Karma is also known to be as the cause and effect law however, I like to call karma the reward and punishing law. According to the law of attraction, we attract what we project out there. We project positive energy when we make ourselves and people around us happy, we strive to bring smiles and rejoice our friends and family. This energy is released out there in the form of vibrations or in other forms such as transactions, physical forces, etc. yet it causes a reaction for the people who either benefit from it or are hurt from it. The law of karma works without distinction and discrimination. Once you release that energy from your mind, body or other resources knowing that it will impact someone in a certain way, it's too late to take that back. That energy magnifies and if it is highly damaging, it will expand and make the initiator a great part of it.

Many of us, - or at least those who hold high morals and standards follow the teaching Confucius (circa 500 B.C.), who had claimed, "Do unto others as you would have done unto you." Aren't we afraid that if we hurt, especially spiritually, the law of karma will strike back at us with a vengeance tenfold? Well, this explains our human nature that, we are governed by a higher power that sets the rules and boundaries for humanity to continue to advance. Imagine if no one practiced the law of karma. The world will be in total chaos. Not that it is not, but in a more sophisticated and subtle way. Many phenomena, we cannot control, they are set in place for our preservation. Nor can we control the behavior of others, but one thing we have total control over, is our behavior and by controlling our behavior we can govern the law of karma, instead of the law governing us.

The doctrine of karma is not taught in modern countries where their education system is based on secularism. Parents in modern societies do not teach doctrines that have something to do with things outside of their culture. Nevertheless, more and more western societies are embracing the karmic law, but on an individual level. Karma is just a word thrown around many times as a way to remind people that we have a defense mechanism. In other words when we say, *karma will get you* we mean, do not hurt me. This was taught throughout the most ancient regions of the world, from Egypt and Babylon to Greece and Judea. The oldest spiritual scriptures come from India and were written long before the Bible was ever conceived, while "karma" derives from the Sanskrit word kri, meaning "to do." Consequently, all action is karma, although traditionally it has most often been associated with the effects of our actions. What the Hindu Vedas (the earliest Indian scriptures) reveal is that, "Our soul is reborn into new bodies in accordance with its former works." The Vedas were also the first to proclaim, "Whatever deed he does, that he will reap." It would be many centuries later that the writers of the Bible would borrow this ancient wisdom. In the 'Bhagavad Gita,' we are told that producing karma in this world is about as inescapable as breathing.

Healing the soul: The art of meditation

People who practice mindfulness have made meditation a daily routine to de-stress while others – more advanced practitioners use meditation to get in touch with the divine and use it as guidance, wisdom, and a tool to make decisions.

These practitioners are highly skilled –usually psychics, healers, gurus, pastors etc. However, meditation is not a monopoly of the paranormal. Everyone can meditate and foresee important events that had happened or are currently happening. I've found six elements of meditation to be crucial to enter the realm of the intelligent world and achieve success. By success I mean practicing healing. Meditation is an art, and as such it requires practice. The more you practice the better you'll get at it.

The main purpose of meditation is to pay attention to what is happening in each moment in the mind. So, when you sit still for meditation, naturally the thought will first interfere, because the thoughts in us cannot stay still.

Without fighting them, bring your attention to the thoughts without getting involved with the thoughts. Meaning you are watching your thoughts as an outsider, by taking a step back.

You can use breathing as a tool to bring your attention to the thoughts without drifting away. When there is such awareness of your thoughts, we can say that you are grounded and ready for meditation or you are in meditation. (Madhu Sai Deevanapalli, 2021)

Element 1: Setting a goal or a purpose for the meditation

There are many types of meditation, and people meditate in various places and in various ways. There are no limits in meditation, and as I'd explained in the previous chapters, one can mediate in quiet or loud places if he or she feels comfortable. Yet the most popular and talked about are: Practicing mindfulness meditation, transcendental meditation and chakra meditation. I practice three of them according to the occasion, yet every one of them has a specific purpose. For example, while mindfulness meditation can be a gateway to the meditation world for debutants, transcendental is for those who seek to get closer to the divine and chakra would be ideal for healing and working with energy.

Element 2 finding a quiet place, preferably undisturbed

Finding a quiet place to meditate is key for achieving mindfulness. Some people take meditation very seriously and have their sanctuary set up inside of

122

the house where people sometimes decorate with icons, altars and make it very spiritual. Some others prefer to meditate in nature. For example, I live near the beach and I have found that it is one of the best places to meditate especially when it is empty. Other people like to meditate at a park, near a lake and so many places where they can find peace and tranquility. It is very important that you remain undisturbed.

Element 3 setting intention focusing on your inner self

Meditation is an art and at the same time is a cognitive demanding process of higher intelligence. The context in meditation is drastically reduced and through the power of the mind, you are the one to create your own context. Putting your psyche to work, requires focus and energy. It requires mental preparedness. Setting an intention and concentrating on the inner self is one of the most common practices. For example, basic meditation entails silencing your mind, relaxing, de-stressing etc. A more sophisticated intention would be visualizing and healing etc. While we meditate and set intentions, we focus primarily on ourselves, unless we are doing spiritual work for others. During this step, it is important to prepare yourself for the upcoming phase such as being aware of the surroundings, a good posture, body and physical readiness. This step also demands a mental readiness. It is the moment where you become aware that for the next twenty minutes or so, nothing else exists but you. Be aware of your entire being and start appreciating everything within you. In every meditation session, you will embark on a very important mission. Things that are revealed during meditation come from above or the Universal consciousness and provide a glimpse of what's out there only happens with meditation.

Element 4 breathing

Our breathing is one of the most vital processes in our body. Thanks to breathing our body receives oxygen that enriches the blood and is then fed to the brain. Thus, I cannot emphasize enough the role that breathing plays in the wellbeing of a person. Breathing has a very unique meaning, in old Scriptures of Hinduism breathing is associated with the energy and that essence is called Prana.

Prana is none other but that energy that gives life to our subtle body. Regulating "Prana" brings equilibrium to the body and mind. Our breath goes to the brain and from the brain it disperses throughout the entire nervous system. To practice meditation, breathing is used in a few ways. Primarily pranic breathing is used to fulfil the

124

physical need for oxygen, so the body is replenished. [87]Secondly, by inhaling and exhaling, your subtle body receives the energy that comes with breathing. This process is also known as bringing life into your body, giving your entire being life and vitality. Thirdly, the pranic breath serves as a means of detachment to cleanse your mind from impurities, attachments, memories, emotions, whether good or bad. It helps your mind to focus and concentrate.

Tip:

> *To experience breathing, sit in a comfortable position. Close your eyes and take deep breaths from the diaphragm. Inhale slowly and exhale in the same way. Feel your belly go up and down as you inhale and exhale. Repeat for a few minutes or for as long as you are comfortable.*
>
> *Expand your lungs: Start with yogic breathing or pranic breathing. As you breathe in, hold your breath for a few seconds (5-10) seconds. Repeat a few times a day for a minute or two.*
>
> *While starting meditation, start with breathing. Do not breathe heavily and abnormally. The breathing process should be as normal as*

---

[87] SIDHU, T. (2020). *Meditate: Breathe into meditation and awaken your potential.* Canada: Terry's Meditation Studio.

possible. If performing regular meditation, individuals with breathing difficulties should breathe normal, and be aware of their breathing. Those who suffer from asthma and other respiratory illnesses, this is a great breathing exercise. Get in a very comfortable position where you are not bothered by the environment and close your eyes. As you breathe in and out notice where your obstruction is. Visualize prana flowing inside of your airways going through your body and bursting it with yogic energy.

Once you are finished with your meditation, do not stop abruptly, bust slowly open your eyes.

Element 5 detachment-

Detachment from reality is an important part of meditation. We often live in a world where we are involved and vested materially, emotionally, and physically. We have spouses, children, parents, friends, family co-workers that we interact with. With social media, many things have become more complex. Social media has shown a side that many of us in the physical world would have a hard time revealing, yet in social media, ruthlessness has become the norm in many groups. Are we that cruel in real life? Many of these virtual interactions have caused hurt, complexity, vulnerability and heartache. Our lives are not as easy as they used to be. Parallel to this cruel reality, there is another reality where we chose to detach from everything that keeps us hostage. That is called detachment. To detach is not easy either. Even when people close their eye and start the meditation process, the flow of thoughts come and go. Many of these thoughts are stored in our subconscious and we do not have control over it. The thoughts most of the time come in the form of emotions and while the negative emotion controls most of your thinking, then it defeats the purpose of healing inner self, or even healing physically. [88]

Tip

To detach from reality, sit in meditation. While you ask for an insightful, enlightened meditation you can experience the feeling of oneness, and nothingness. Try to clear your mind, as I had written in the previous chapters. If your mind gets invaded by thoughts, come back to breathing, and start quieting your mind. Many practitioners in the meditation reassure that with time, people will get used to quieting their mind. Other techniques include counting to keep your mind away

---

[88] SIDHU, T. (2020). *Meditate: Breathe into meditation and awaken your potential*. Canada: Terry's Meditation Studio.

from distraction. Once you are detached from the reality, worries and concerns then meditation must proceed mindfully and quietly. It does not matter what you would like to achieve in the meditation, as long as you meditate.

Element 6 letting go of every thought and experience the divine

Let go of everything, breathe, and experience the moment. Get lost in the beauty of the divine. In this stage of the meditation, whether regular or transcendental, letting go of all your worries is key in attaining full awareness of your inner self. It is one step closer to understanding that your inner self belongs to the higher self- higher power. It should be a moment that leads you realize the Atman – the soul. [89]

Tip

In a quiet place sit in a meditation position, or however makes you feel comfortable, allowing the source, or the holy to give you the answers to your questions. Thus, there are three prerequisites to achieve a mindful experience and meditation.

1. *Logistics*
2. *Emotional preparedness*

---

[89] SIDHU, T. (2020). *Meditate: Breathe into meditation and awaken your potential.* Canada: Terry's Meditation Studio.

3. *Physical preparedness*

4. *Spiritual preparedness*

Meditation is the art of getting in touch with your inner self, your spirit guide, God himself or other entities. Meditation can be performed in many ways, here below I have a list of various types of meditation:

Meditation and cleansing your mind of impurities:

Preparing your body for meditation

People take meditation for granted. Meditation is a process where you get in touch with the higher power and as such it needs special attention. Your body must be prepared before meditation spiritually and physically. What we put in our body reveals our physical health. Meditation must be a daily practice or a few times a week at least yet your body needs to be ready. Our lives are busy and when you read this, you'll probably evaluate your options. *I don't have time, I work, I have children* and so forth. Even if it takes five minutes in your room, it would be beneficial. The best meditation results are in the morning upon waking up. Your stomach must be empty. Or if you decide to put something in your stomach, let it be water, or warm water with a teaspoon of honey.

It is not beneficial to eat before you meditate, for the simple reason that when we eat, the blood rushes in the stomach to digest the food. When people eat it will take at least a few hours to feel empty. The breathing won't be the greatest on a full stomach.

Many people who practice meditation are vegan or vegetarian. To meditate you do not have to give up meat, chicken or fish. Just make sure that you meditate before you eat or at least three hours after you eat. I do not recommend changing your diet, but if you choose to cleanse your body, you can eat vegan/vegetarian for a few days to increase your vibration, and you get back on your usual diet after that. You will experience better meditation sessions. The third eye chakra will open much more easily and efficiently. Soft drinks are perfectly fine. Alcohol is not recommended before meditation or in the system during meditation. It will impair the ability to have a divine connection and it makes people vulnerable to low vibe frequencies. Thus, in conclusion, the best meditation is when people fast –they give up meat, chicken, fish and eggs, for one day or for a few days. This makes the path to the Universal Consciousness much clearer.

Sanctuary

If you desire to have a rich spiritual life, you must build your sanctuary. A sanctuary is a place where you can be away from others, where you have your altar, candles, herbs, incents, anything that helps you in your prayer or spiritual work. There is a perception that we must not pray to idols, and icons. In fact, Hinduism and Buddhism have an abundance of statues, temples, figurines and other cult objects. Religions and philosophies come in many forms and beliefs. Cultures pray according to their rituals and believe the idols carry or transmit the energy. Meaning, the statues serve as an outlet, and a transmitter of God's positive energy. When performing meditation people can use various cult objects, from icons, figurines of God, prophets, avatars etc. Every sacred object is very beneficial during meditation because it attracts positive vibes and can enhance the receiving power. For best results, when meditating, you can light a candle or purify the air with incents without inhaling the smoke, because that can be problematic for people who have respiratory issues.

Once you set your altar, you are ready to perform a powerful meditation, or a powerful healing meditation.

People who are used to meditating, can do so without being bothered by the surrounding environment. However, meditation is the time to get in touch with the higher power to get into the Universal Consciousness, therefore, being alone in your sanctuary is very important.

Meditation enhancers

Crystals- a crystal guide or book can help perform healing meditation by either using crystal grids or candle burning.

Herbs- can help purify the environment, bring positive energy, and cleanse the house. Also, herbs sold at the local market such as herbal teas help for the purpose that they are used.

Oils – although, burning oils should be used with caution because they release chemicals while burning, essential oils can be beneficial for treating various areas of the body and various conditions such as arthritis. People use essential oils before meditating but depending on the person, they should be used with caution. Peppermint, lavender, citrus orange are great oils to open up the intuition. Always use as directed in the bottle or by the vendor.

When performing meditation people can use various cult objects, from icons, figurines of God, prophets, avatars etc. Every sacred object is very beneficial during meditation because it attracts positive vibes and can enhance the receiving power.

Chakra meditation

Chakras are subtle energy wheels in our subtle body. When you read the chapter of quantum physics and learn about the presence of energy in our bodies – in a more scientific way, the chakra system is the same concept, yet in a more traditional way. It has been practiced for thousands of years mainly by Hindus used in healing practices. Eastern medicine differs drastically from western medicine because it uses a more organic approach, close to a natural way of healing. It is very important to understand the subtle energy and how that operates. When we say energy wheel, what do we mean by that? Chakra comes from old Sanskrit –and means exactly wheel. Every author who talks about meditation, cannot skip the subject of the chakra and how we relate to it. Chakras cannot be felt or touched, yet they are part of the subtle body.

In the old East Indian Tantric Scriptures, there are known to be seven chakras. They align in the spinal cord, start from the lower end of the spinal cord until the crown or the top of the head.

These energy centers in the body are called chakras. There are one hundred fourteen chakras, all with different points all over the body. They are mainly responsible for our internal exchange of energy. Yet, the following seven main chakras are particularly important: Root chakra, sacral chakra, solar plexus chakra, heart chakra, throat chakra, third eye chakra and the crown chakra.[90]

While we perform chakra meditation, we must use at least ten to fifteen minutes to cleanse and reenergize the chakras. The best way to meditate on the chakras are listening to crystal bowl or Himalayan gongs. They produce sounds that can easily activate the chakras. While doing a routine chakra cleansing for fifteen minutes it is beneficial to find a very quiet place where you will not be disturbed. For best results hold crystals in both hands. You can get into a yoga position if you are comfortable, but if you have back problems you can also lay down. A sitting position can give optimal results: Before I explain guided meditation, I would like to go over the chakras and their properties:

---

[90] 7-Chakras-for-Beginners-The-Complete-Guide.pdf Mindmonia. (n.d.). *7 chakras for beginners the complete guide* [Brochure]. Author.

Chakra Image [91]

Chakra properties
Root → Base of the spine, survival safety, red
Sacral→ Lower abdomen, creativity sexuality, orange
Solar plexus→ navel center, identity, yellow
Heart→ Chest and heart, love, compassion, green
Throat→ Throat, communication, blue
Third Eye → Between eyebrows, intuition, indigo
Crown → Top of the head, coaction with the higher power.

While the three first chakras are related to our physical body, the last three have to do with our

---

[91] Royalty free image

spirituality. Whereas the heart chakra belongs to our physical body because it's in charge of two main organs: Heart and lungs – on the other hand, the heart chakra dwells in a spiritual realm and is the chakra that shelters the most divine feeling: Love. [92]

Chakra meditation procedure:

Sit in a quiet place

Find meditation music that will enhance your need to meditate. If you are doing a chakra cleansing with Himalayan gongs, make sure you download the track so the commercial does not interrupt your mood, if not you can perform the chakra cleansing on your own. Start with the breathing technique. Breathe in and out. After performing breathing and relaxation for two or three minutes, start visualizing the chakras as a wheel of colors. Start with the red chakra and work your way up.

After chakras are restored, picture a bright light that goes through your body from the crown to the sacrum. Picture light bursting out and filling your body with radiant and vibrant colors. You can also visualize the white light to cleanse your body of negative energy. After you finish the meditation, you will feel drowsy and very relaxed. You can do the

---

[92] SIDHU, T. (2020). *Meditate: Breathe into meditation and awaken your potential.* Canada: Terry's Meditation Studio.

chakra tuning up anytime during the day. If done at night, it will help you get a good night's sleep.

Asking questions

Trust your god, trust the spirit. Whatever you hear or see in this meditation technique comes from the source.

We ache because it's in our human nature to ache. We hurt because we have feelings and our ego is put to the test. We hurt for many reasons, and when we do, we seek answers. We cry often and when we do, we ask our god *why?* We get lost and confused when we cannot explain what is happening to us. It is in our human nature to ask for closure in explanation to events in our physical world. Using our senses and our mind, sometimes the mind deceives. Rarely, people say, *I will get into meditation and ask God for answers.* It's as simple as that. The answers are right in front of us, we just have to learn how to ask, how to receive and to trust what comes from the above as the truth. Meditation is just like the other skills, the more you practice, the better you'll be. Taking into consideration the fact that we have busy lives, we cannot afford to meditate and retreat for hours, yet fifteen minutes a day, is more than enough to practice mindful meditation.

When people are in distress, they look for an answer to their problem. Often people find themselves in a deep emotional turmoil. When gurus say that you have to be calm and happy, it sounds as if they've never experienced distress. We cannot deprive ourselves from certain emotions and feelings, because human nature does not work like that. We cannot have a bad day at work and be happy because being sad will bring us more sadness. It is important to have a sane mind and high spirit, but sometimes it's quite impossible especially when we experience abuse, depreciation, harassment and all the feelings that others project on us, making us feel worthless. This will not prohibit us from achieving great spiritual work through meditation. In fact, the messages we get from the source, when we experience strong emotions are quite accurate. For example, when you need to know what happened beyond what we see and hear, there is a lot more information out there that we can receive from the source when we ask for clues.

How does that work? You must ask specific questions.

The answers will not come easily. This technique requires skills because it has to do with going to the source. Once at the source, the answers might come in various ways. They can come in the form of visuals, images, words, sounds or feelings. The most common are words, sentences and images. The source will allow you to receive answers, and you have to trust the answers and not question the veracity. These answers serve to make you aware and be cautious, but do not act irrationally upon them. Be aware of everything and your surroundings. Spirit guides do not like to be dismissed, they work hard to get you to the source and give you the answers. Depending on your questions, the source is pretty good in giving you answers that, you might or might not even expect. Sometimes the source will provide insight into getting closure on something that might have bothered you. These answers might be ambiguous at times and might confuse you. In that case, ask the source for guidance, and it will come to you in many forms, sometimes even peculiar, but we have to realize that the source holds tremendous amounts of information and the truth.

How to get to the source

There are a few ways of how to get to the source. However, people have to understand that getting to the source is not something that happens in the physical world. We use our physical mind, and our physical senses to gather information to create a connection with the source. Yet this requires energy and tremendous focus. In addition, to get to the source one must enter a deep meditation through self-hypnosis, or just through concentration.

Nothing should be forced. Images should not be forced in this type of meditation. While images are great for healing and other spiritual work, getting to the source will happen naturally. The spirit, or the guardian angel is in charge for bringing people to the source.

Spirit Guides are like your well-wishers who are trying to help you when you need them. See them as your new friends. Slowly build relationships with them. Imagine you are traveling on an airplane for a long flight, what would you do with the person sitting next to you? You try to get to know them, share, listen, ask questions, and so on. Then when you develop some kind of bond, you try to open up and connect more. Initially you may be reserved and hesitate but when you get to know them better, you could build relationships. Similarly, spirit guides are actually part of you, they are energies that connect through your senses, your five elements, your thoughts, in a particular form or shape or color. Talk to them and start making friends. There is nothing to be afraid of. Your fear is not allowing you to connect with them. Don't listen to your fear or let your fear control you. (Madhu Sai Deevanapalli)

The source is otherwise known as the Universal Consciousness and every person has a different mental image of the Universal Consciousness.

To get to the source one must be ready, one must be willing, and patient. While the guiding spirit takes us to the source, we must let go of everything, every thought, and every worry. Our subconscious, mental images, emotional thinking and all other influences that might impair the connection –must be separated from the process.

The body should be completely relaxed. In case your mind wanders, bring the attention back to nothingness. Once nothingness is achieved, wait for the Universal Consciousness to appear. To some people the Universal Consciousness will appear as the mental image of the Eye of God. To others, it might appear like a beautiful beach paradise, or a luscious forest. It does not really matter. All that matters, is we enter the realm and the vibrations of the source to experience bliss or to seek the answers that we look for.

There are thousands or writers and practitioners in the field and everyone has their uniqueness, their perspective in explaining what the source is. Some emphasize the importance of being in alignment with the source completely overlooking the human nature of people.

Once we are in the source, the experiences are countless. If the meditation requires answers, you will receive answers, sometimes directly or through symbolism. Spirit is very smart, and that is how we explain the parables in the Bible, or the poetry in the Vedas, or the teaching of Quran. Gods sometimes do not speak directly. They like to surprise us with endless possibilities to seek answers. Nevertheless, the answers can be direct.

Visualizing

This is a technique used in healing, but also seeking answers. In this section I will focus on healing or changing things positively. People have to realize that the positive science works with positive mental images and helps push them in the right direction. We as human beings sometimes are egoistic and as such, we project our ego to the rightest. But that is not necessarily true. Sometimes our happiness can be someone's distress and forcing mental images on something or someone when we know will make another person unhappy is useless. It will not happen, even if it does, we will be unhappy. Because if the person is forcing something against beliefs or another person's will. This is against the positive science.

Visualizing for healing can be done directly by creating images of healing whereas an ailment, a broken heart, a physical wound, an emotional state of being.[93]

Hypnosis (self-hypnosis)

Hypnosis is a form of deep meditation. It is a method of instructing your conscious mind to relax, and stop thinking, just like you behave in your sleep, but without actually falling asleep. [94] Hewitt suggests that once your conscious mind relaxes, then the subconscious mind takes over and becomes receptive to suggestions. In fact, in earlier chapters we learned that the subconscious mind is not the one that reasons, but it is the one that listens and follows just like an obedient "servant." In order for it to work, you must believe in your ability to achieve mindfulness and results. Self-hypnosis is a constant affirmation of your desires and a positive reinforcement of your goals said in a deep meditative

---

[93] Disclaimer: A healing visualization cannot replace medical advice or therapy done by a medical doctor.

[94] Burke, A. (2004). *Self-hypnosis: New tools for deep and lasting transformation*. Berkely, CA: Crossing Press.

state. From a practical point of view, everyone can perform a hypnosis session just by writing the script. There are many ways of performing hypnosis. The most successful way is a scripted, recorded affirmation.

Work with energy: Astral meditation

The purpose of Astrology is to promote self-realization[95]

The planets are humongous celestial bodies and they embody tremendous energy. Planets are bound with one another pulling and pushing using the same energies. For instance, the high tide and low tide is caused by the effect of the moon and the earth pulling and pushing. The lunar cycle has a great effect on our subtle bodies and physical bodies. The major effect of this cycle is felt during the new moon and full moon. They both impact our moods, our breathing, and our wellbeing whether we like it or not. While the moon is friendly to some signs it can cause uneasiness to others. For example, the moon is not friendly to me at all. During the new moon I have difficulty breathing and since I had respiratory issues, the night of the new moon and the full moon,

---

[95] Fish, R., & Kurzak, R. (2012). *The Art and Science of Vedic Astrology (The Foundation Course Book.* Asheville, North Carolina: Asheville Vedic Astrology.

it felt as if a truck sat on my chest. The next day it was back to normal. Also, during mercury retrograde, effects of the communication disasters are felt everywhere. I must say, it is a dark period in communication, contract signing and relationships. Each of us have a ruling planet and of course not so friendly planets. The most important thing is to know how to protect ourselves from the astral energy or use it to our benefit.

CONCLUSION

In this book we discussed various arguments that dealt with self. First, I argued that we have two bodies, a physical body and a spiritual body of light. I explained that we are social beings and as such our understanding of identity starts with gender, then the social group where we think we belong. I discussed that individualism and individual freedoms have always been under the assault of institutionalized religion and corporate psychology and mentality. Then I explain what group identity is and how we are trained to think collectively. I then, give an overview of the concept of self and inner self in Eastern philosophy with Buddhism and Hinduism and Western Philosophy, Socrates and Descartes. In the following chapters I explained what the subtle body is and how we understand the soul as it pertains to our inner self. I give an overview of the energy presence in our physical body and how matter and energy are two forces that are bound under the principles of quantum physics. In the chapter of mind and thought, I explain how the theory of mind works and how people make use of their thought process and mind power to reach out to Universal Consciousness. I contrasted three main concepts, the conscious mind, the subconscious mind and conscience. These three concepts have to do with a process in our brain that enables us to make that connection between our inner self and the Universal Consciousness. I also touch upon the concept of Universal Mind and explain the source of knowledge, power and pure essence. Finally, I explain what meditation is as a process and I gave a few meditation and spiritual practices for people to enrich their lives. I had decided to write this book because I wanted to improve my life, to expand my Universal Consciousness, opportunities, and in this book, I explained exactly how I successfully live a fulfilling life without headaches, problems, and with harmony within me.

Well, I was not always happy and enthusiastic. There were times when I'd felt anxious and anxiety would take over my life. I have a strong mind, but not such a strong physique. Meaning, I am not that strong physically. My stress always translated to physical pain, whereas back ache, shortness of breath overeating, becoming overweight and so forth. I was and still am very spiritual, and whenever I need answers, I know what to do. Luckily in life I always used rationalism and logic, but I've made a decision based on my meditation practices. I have encountered some mortal enemies out there. Sometimes logic and rationalism might work with you and inert objects, but when it comes to dealing with people, and especially enemies, you have to have a different approach. Sometimes we feel powerless, but in fact we are not. Once I started practicing meditation, I found the answers that I wanted, when I needed them.

Generally speaking, I was always a shy person, I wanted to be known for someone, smart and beautiful. I have been put down by many people. Still somehow, I always come out victorious with a personal and professional plan. God has always taken me out of harm's way – literally.

The feeling of unhappiness, what causes it?

Happiness and unhappiness are opposite feelings. Humans are social "beings," and as such, they are heavily impacted by the surrounding environment.

Unhappiness is a deep feeling of sorrow. It is the regret for things that we could not achieve or were out of our reach. Unhappiness is an emotional state that is caused by various factors, environmental and biological. I want to focus on the immediate environment because it is very accessible to us, and certain things can be changed with meditation. If we cannot change things, we could walk away when we least expect it. I am hoping that this book will allow a different view of self where we all see it differently and do not take it for granted. This book will also serve as a bridge for me to connect to the higher power. As I sit in meditation and ask for answers, I cannot help but reveal, that I am trained to critically write mainly in political science and leadership. That is how my brain works, yet I believe that shifting my awareness to what people love to hear and read, gives me strength to help by writing about spirituality instead of politics. Till the next book, I wish you a beautiful spiritual journey.

Love

Dr. K

BIBLIOGRAPHY

Achuthananda, S. (2013). *Many, many, many Gods of Hinduism: Turning believers into non-believers and non-believers into believers.* Charleston, SC: CreateSpace Independent Publishing Platform.

Bailey, A. A. (1931). The soul and its mechanism. *The Journal of Nervous and   Mental Disease, 74*(1), 123. doi:10.1097/00005053-193107000-00078

Burke, A. (2004). *Self-hypnosis: New tools for deep and lasting transformation.* Berkely, CA: Crossing Press.

Burke, P. J., & Stets, J. E. (2009). *Identity theory.* Oxford: Oxford University Press.

Chaffee, J. (2015). Philosopher's way: Thinking critically about profound ideas. In *Philosopher's Way: Thinking Critically about Profound Ideas* (pp. 92-104). Pearson.

Christmas Humphreys (2012). *Exploring Buddhism.* Routledge. pp. 42–43

Chung, Sung Jang. "The Science of Self, Mind and ody." *Open Journal of Philosophy* 02, no. 03 (2012): 171-78. doi:10.4236/ojpp.2012.23026.

David Lorenzen (2004), The Hindu World (Editors: Sushil Mittal and Gene Thursby), Routledge, pages 208-209

Deussen, Paul and Geden, A. S. The Philosophy of the Upanishads. Cosimo   Classics (June 1, 2010). P. 86.

E = mc2. (n.d.). Retrieved April 18, 2021, from https://www.britannica.com/science/E-mc2-equation?utm_campaign=b-extension&utm_medium=chrome&utm_source=ebin sights&utm_content=E+

Etheric body - soundessence.net. Retrieved April 18, 2021, from
      https://www.soundessence.net/assets/etheric_bod
y.pdf

James 3:16 ESV - - Bible Gateway. Retrieved April 15, 2021.

Karpouzos, Alexis. *Universal Consciousness: The Bridges between Science and Spirituality*. Athens, Greece: Think.Lab, 2020.

Kieffer, S. (2018, December 03). How the brain Works: Johns Hopkins Comprehensive Brain Tumor Center.

Mayfield brain & Spine. (n.d.). Retrieved April 19, 2021, from https://mayfieldclinic.com/pe-anatbrain.htm

Mindmonia. (n.d.). *7 chakras for beginners the complete guide* [Brochure]. Author.

Plato, Jowett, B., & Edman, I. (1928). *The works of Plato*. New York: Simon and Schuster.

Rahula, W., & Demiéville, P. (1962). *What Buddha Taught*. New York: Grove Press.

Richard Gombrich (2006). *Theravada Buddhism*. Routledge. p. 47

Robinson, H. (2020, September 11). Dualism. Retrieved April 17, 2021.

Romans 2:8 ESV - - Bible Gateway. (n.d.). Retrieved April 15, 2021.

Ruhl, C. (2020, August 07). Theory of mind. Retrieved April 20, 2021, from https://www.simplypsychology.org/theory-of-mind.html

Sasson, R. (2021, April 17). Concentration exercises for training and focusing the mind.

Segrest, D., & *, N. (2020, September). Archetypes and symbols. Retrieved April 18, 2021.

Shumsky, S. G. (1996). *Divine Revelation*. New York: Fireside.

SIDHU, T. (2020). *Meditate: Breathe into meditation and awaken your potential*. Canada: Terry's Meditation Studio.

Stone, V. E., Baron-Cohen, S., & Knight, R. T. (1998). Frontal lobe contributions to theory of mind. *Journal of Cognitive Neuroscience, 10*(5), 640-656.

The Astral and Causal Bodies. http://www.simonheather.co.uk/pages/articles/the_etheri c_body.pdf

Vedantany1894. (2018, December 19). || the atman || by Swami Sarvapriyananda. Retrieved   April 15, 2021,

Vedantany1894 (Director). (2021, January 14). *Ask Swami with Swami Sarvapriyananda | January 3RD, 2021*

ULE, A. (2016). The concept of self in Buddhism and Brahmanism: Some remarks. *Asian Studies, 4*(1), 81-95.

What is the causal body? - Definition from Yogapedia. Retrieved April 15, 2021.

45 Metaphysical Foundations of Modern Physical Science, p. 275.1

152

Manufactured by Amazon.ca
Bolton, ON